Mammalian Artificial Chromosomes

METHODS IN MOLECULAR BIOLOGY™

John M. Walker, SERIES EDITOR

METHODS IN MOLECULAR BIOLOGY™

Mammalian Artificial Chromosomes

Methods and Protocols

Edited by

Vittorio Sgaramella

and

Sandro Eridani

Istituto di Tecnologie Biomediche Avanzate, Consiglio Nationale Ricerche, Segrate, Italy

Humana Press ✳ Totowa, New Jersey

Preface

In 1996, we organized a workshop, *inter alia*, at the National Research Council in Milan under the generous sponsorship of the European Science Foundation. On that occasion, a small group of investigators convened from many countries and presented early evidence of the possibility of assembling basic units of mammalian chromosomes into artificial constructs (or, indeed, reducing the relevant components to more manageable dimensions and defined constitution).

Progress in the following years has been slow but steady. Many scientists who took part in the workshop have since been engaged in active and productive research. It goes to the credit of Humana Press to have realized the need for a book on artificial chromosomes that aims to provide better tools to all scientists committed to this field who are confronted with very difficult technical problems.

We have strived to cover in *Mammalian Artificial Chromosomes: Methods and Protocols* all relevant areas of artificial chromosome research, from basic genetics to daring attempts to build new tools for genetic therapy. We are of course grateful to the authors who have accepted the task of describing the technical steps and pitfalls that can be encountered in their research. Rarely has a very delicate methodology been presented with such meticulous care.

We have been helped in this enterprise by the excellent librarian of the LITA Institute in Segrate, Italy, Ms. Claudia Piergigli, whom we thank warmly. Ms. Francesca Tarchi, ITB-CNR secretary, was also helpful. Ivo Castagna and Alberto Ribolla, provided useful technical support. Lastly, we thank Professor John Walker and Craig Adams of Humana Press for their patience and understanding during the preparation of this book.

Vittorio Sgaramella
Sandro Eridani

Contents

Contributors

GULNARA ABDURASHIDOVA • *International Centre for Genetic Engineering and Biotechnology, Trieste, Italy*

FIORENTINA ASCENZIONI • *Dipartimento di Biologia Cellulare e dello Sviluppo, Universitá "La Sapienza," Rome, Italy*

ANITA H. BOROWSKI • *Abgenix Biopharma Inc., Vancouver, British Columbia, Canada*

JÉRÔME BUARD • *Centre National de la Recherche Scientifique (CNRS), Montpellier, France*

DEBORAH O. CO • *Chromos Molecular Systems, Inc., Burnaby, British Columbia, Canada*

HOWARD COOKE • *Chromosome Biology Section, MRC Human Genetics Unit, Western General Hospital, Edinburgh, UK*

LEONARDO D'AIUTO • *Chromosome Biology Section, MRC Human Genetics Unit, Western General Hospital, Edinburgh, UK*

JOSE I. DE LAS HERAS • *Chromosome Biology Section, MRC Human Genetics Unit, Western General Hospital, Edinburgh, UK*

ALBERTINA DE SARIO • *Centre National de la Recherche Scientifique (CNRS), Montpellier, France*

PIERLUIGI DONINI • *Dipartimento di Biologia Cellulare e dello Sviluppo, Universitá "La Sapienza," Rome, Italy*

SANDRO ERIDANI • *Istituto di Tecnologie Biomediche Avanzate, Consiglio Nationale Ricerche, Segrate, Italy*

ARTURO FALASCHI • *International Centre for Genetic Engineering and Biotechnology, Trieste, Italy*

CHRISTINE J. FARR • *Department of Genetics, University of Cambridge, Cambridge, UK*

ROBERT FEIL • *Institute of Molecular Genetics, Centre National de la Recherche Scientifique (CNRS), Montpellier, France*

JAN FILIPSKI • *Laboratoire de Biochimie de la Chromatine, Institute Jacques Monod, Universitiés Paris 6 and Paris 7, Paris, France*

PIER ASSUNTA FRADIANA • *Dipartimento di Biologia Cellulare e dello Sviluppo, Universitá "La Sapienza," Rome, Italy*

YUJI GOTO • *Insitute of Molecular Genetics, Centre National de la Recherche Scientifique (CNRS), Montpellier, France*

CHRISTOPH GRUNAU • *Centre National de la Recherche Scientifique (CNRS), Montpellier, France*

GYULA HADLACZKY • *Institute of Genetics, Biological Research Center, Hungarian Academy of Sciences, Szeged, Hungary; and Chromos Molecular Systems Inc., Burnaby, British Columbia, Canada*

PETER J. HORNSBY • *Department of Physiology and Sam and Ann Barshop Center for Longevity and Aging Studies, University of Texas Health Science Center, San Antonio, TX*

LEOPOLDO IANNUZZI • *Laboratory of Animal Cytogenetics and Gene Mapping, National Research Council, ISPAAM, Naples, Italy*

PANAYIOTIS A. IOANNOU • *Cell and Gene Therapy (CAGT) Research Group, The Murdoch Children's Research Institute, Royal Children's Hospital, Parkville, Melbourne, Australia*

GRZEGORZ IRA • *Laboratory of Genetics, N. Copernicus University, Toruñ, Poland; and Rosenstiel Basic Medical Sciences Research Center and Department of Biology, Brandeis University, Waltham, MA*

ISAO ISHIDA • *Pharmaceutical Research Laboratory, Kirin Brewery Co. Ltd., Gunma, Japan*

DUANGPORN JAMSAI • *Cell and Gene Therapy (CAGT) Research Group, The Murdoch Children's Research Institute, Royal Children's Hospital, Parkville, Melbourne, Australia*

FRANK R. JIRIK • *Department of Biochemistry and Molecular Biology, University of Calgary, Calgary, Alberta, Canada*

JACEK KRÓL • *Laboratory of Cancer Genetics, Institute of Bioorganic Chemistry, Polish Academy of Science, Poznañ, Poland*

YOSHIMI KUROIWA • *Pharmaceutical Research Laboratory, Kirin Brewery Co. Ltd., Gunma, Japan*

JOSEPHINE D. LEUNG • *Chromos Molecular Systems Inc., Burnaby, British Columbia, Canada*

HAN N. LIM • *Department of Genetics, University of Cambridge, Cambridge, UK*

SAMUEL MCLENACHAN • *Cell and Gene Therapy (CAGT) Research Group, The Murdoch Children's Research Institute, Royal Children's Hospital, Parkville, Melbourne, Australia*

DIANE P. MONTEITH • *Chromos Molecular Systems Inc., Burnaby, British Columbia, Canada*

MICHAEL R. ORFORD • *Cell and Gene Therapy (CAGT) Research Group, The Murdoch Children's Research Institute, Royal Children's Hospital, Parkville, Melbourne, Australia*

CARL F. PEREZ • *Chromos Molecular Systems Inc., Burnaby, British Columbia, Canada*

TÜNDE PRAZNOVSZKY • *Institute of Genetics, Biological Research Center, Hungarian Academy of Sciences, Szeged, Hungary*

JACQUES PUECHBERTY • *Centre National de la Recherche Scientifique (CNRS), Montpellier, France*

GÉRARD ROIZÉS • *Centre National de la Recherche Scientifique (CNRS), Montpellier, France*

VITTORIO SGARAMELLA • *Istituto di Tecnologie Biomediche Avanzate, Consiglio Nationale Ricerche, Segrate, Italy*

KAZUMA TOMIZUKA • *Pharmaceutical Research Laboratory, Kirin Brewery Co. Ltd., Gunma, Japan*

ALESSANDRO VINDIGNI • *International Centre for Genetic Engineering and Biotechnology, Trieste, Italy*

1

From Natural to Artificial Chromosomes

An Overview

Vittorio Sgaramella and Sandro Eridani

1. Introduction

The rationale for building artificial chromosomes (ACs) has been critically reviewed by various authors at different stages of this line of investigation, initiated in the early 1980s.

Two major goals have been generally stressed: first, the possibility of a better understanding of the structure and functions of natural chromosomes; second, the challenges presented by their use as large-capacity gene vectors for DNA cloning, in view of genetic improvement (in animals of commercial relevance) and hopefully of disease correction (in humans): for a review, *see* Willard *(1)*. The main features for an efficient mammalian AC (MAC) are (1) a vectorial capacity up to a few megabases; (2) a manageable size for their in vitro manipulation; (3) a correct intracellular location and copy number; (4) no untoward effect on the host cell; and (5) the ability to express the transgene (or transchromosome) in a physiological way *(2)*.

From: *Methods in Molecular Biology, Vol. 240:*
Mammalian Artificial Chromosomes: Methods and Protocols
Edited by: V. Sgaramella and S. Eridani © Humana Press Inc., Totowa, NJ

A number of problems have arisen in the creation of ACs *(3)*. In the first place, some recently generated minichromosomes (to be considered intermediates in the assembly of AC) do not show the expected relationship to the input DNA: this may be the result of an intrinsic structural instability of the constructs *(4)* that makes them more mutable. Also frequent is the occurrence of a mitotic instability that causes their loss on cell division. Another possible disadvantage of the presence of an AC in a cell might be some interference with resident natural chromosome sorting during the cell division.

The challenges raised by these problems have been met in recent years with variable success, and the present volume is witness of the increasing efforts to build up a new technology, which may lead to the understanding of the largely unknown parameters governing chromosome formation and, eventually, the creation of stable constructs, with an increased size of input DNA and the potential for a physiological regulation of its prospected genetic functions.

The first ACs were assembled in yeast and were built after the identification of the components required for the upgrading of a plasmid into an AC and allowed a stringent confirmation of the roles of the distinct constituent elements. Moreover, yeast artificial chromosomes have been very useful vectors of large chunks of DNA during the early phases of various genome projects, notably of the human one *(5)*.

As reported by various groups, the use of yeast artificial chromosomes, and in particular of those of the second generation containing large, single segments of human DNA and freed of most cocloning *(6)*, has contributed to the identification of five DNAse hypersensitive sites *(7)*. Of special interest appears the study of the chromatin formed in yeast by one of these sites, the CpG island that flanks the G6PD gene: it demonstrates that variations in C+G overall content and/or CpG frequency may influence the DNA structure, thus modulating the chromatin organization. It also appears that the hot spots of recombination located in the human chromosomes remain recombination prone on cloning in yeast. In Chapter 3,

Filipski et al. describe a detailed experimental procedure that could be used for mapping these important sites.

In recent years, an increasing attention has been given to the study of chromosome alterations in domestic animals, which are often quite difficult to detect because most of these chromosomes are acrocentric and similar in size. It was, however, noticed that cytogenetic alterations were responsible in cattle for a reduced fertility rate, so that a study of such modifications was the basis for many attempts to improve the genetic character of bovids *(8)*. The recent availability of chromosome banding techniques, molecular markers, and painting probes has opened the way for a remarkable advance in our knowledge In Chapter 2, Iannuzzi describes the sophisticated methods now in use for the elucidation of the structure of domestic animal chromosomes, with their relevant implications.

An increasingly interesting phenomenon that took place during evolution is the so-called "genomic imprinting"; this process causes some genes to be expressed according to their parental origin, resulting in asymmetry in the function of parental genomes. Imprinting also determines the choice of which X chromosome is to be inactivated in female cells of mammals. Imprinted genes are important in prenatal growth and development, and are also involved in human disease *(9)*. As to the origins of imprinting, it is now established that there are connections between chromatin modification and structure, DNA methylation and imprinting. In Chapter 4, Goto and Feil discuss the possible impact of imprinting on transgenes and ACs, and point out that manipulation of embryo cells culture may disrupt imprinting and can thereby lead to aberrant phenotypes. These alterations could be relevant to the application of methodologies based on transgenes and transchromosomes.

Chromosomal proteins have been the subject of extensive studies from a variety of viewpoints, addressing both historic and nonhistonic proteins, which apparently elicit many different properties. Histone proteins are known to form the structure of the nucleosome, a central complex, around which a double-stranded stretch of DNA is coiled approximately twice. It has been recently recognized that mutations of the genes coding for these proteins, are responsible for the development of severe genetic disorders (like the α-thalassemia/mental retardation syndrome and other forms), often causing alterations of the central nervous system: these conditions are now called "chromatin diseases" *(10)*.

In a different context, a very interesting family of nonhistonic proteins are cumulatively described as the high-mobility group box proteins, which are important architectural factors for the assembly of DNA protein complexes and their positioning at their binding sites. Beside a nuclear function, however, they seem to possess other activities, like the capacity, when released by necrotic cells in the medium as soluble molecules, to act as signals of cell death, triggering the inflammatory process *(11)*.

Among the basic components of chromosomes, telomeres are so far possibly the best understood. Telomeric DNA is important for the replication, integrity, and independence of linear chromosomes. Particular attention has been devoted to subtelomeric regions, which were believed until recently to represent merely a buffer between the extreme terminal sequences (needed to protect chromosome ends from degradation and recombination) and the essential internal sequences. However, many subtelomeric regions have revealed a high content of genes and are now considered functional parts of the expressed genome *(12)*.

The formation of telomeres at the MAC termini requires the presence of at least some wild-type telomeric repeats, which function as "seeds" or primers for polymerizing enzymes: those repeated sequences seem to match the binding sites of a short RNA component (chromosomically coded) of a candidate telomeric ribonucleoprotein complex, which in synergy with other factors is necessary in mammalian cells for telomere formation and completion. Chromosomal DNA ends tend to shorten gradually at each DNA replication cycle, because of the fixed 5'–3' direction of DNA synthesis: such effect can be overcome by telomerase, a reverse transcriptase that forms telomeric repeats DNA at the telomere 3' terminus, using the RNA segment present in the telomerase as template. It is of interest that in cells where a forced expression of the reverse transcriptase component of telomeric ribonucleo-protein complex is achieved, the progressive shortening of telomeres is prevented. Moreover, clones with this property become immortalized and show optimal survival and function when xenotransplanted (*see* Chapter 8 by P. Hornsby).

The implications of telomerase activity for survival and function of the telomeres are discussed in Chapter 7 by Ascenzioni and coworkers, with a detailed illustration of methods to test telomerase activity, like the telomeric repeat amplification protocol assay, which is widely used as a tool to evaluate tumor progression and to determine the efficacy of therapeutic interventions. The telomerase, which is overexpressed in human tumors and seems to be essential for cell immortalization, has become a major target for therapeutically promising studies in swift progress along this direction *(13)*.

Another essential component of chromosomes is the centromere, a repetitive DNA sequence that is involved in the preliminary pairing and subsequent segregation of chromosomes to daughter cells during cell division. The centromere, however, may be either localized or dispersed along the chromosome but is still capable, in the latter case, of properly functioning as required *(14,15)*. These structures, called neocentromeres, have attracted considerable attention; Roizés and coworkers review this topic in Chapter 5, discussing not only the DNA sequences involved, which may be unrelated to the canonical sequences of the old centromeres but still able to exert centromere functions. In humans, one of these epigenetic factors is the CENP-A protein, which is thought to play a central role in the process of centromerization because it shows a high affinity for the centromeric DNA sequence. Other proteins, including non-H3 histones, may be involved in the building up of a centromeric structure; DNA methylation is also considered an important factor in the induction of centromeric activity. Specification of a locus to become a centromere can therefore be attributed to the concomitance of different factors, which would ensure its activity through many generations.

An interesting issue is the minimal sequence requirement for proper centromere function and chromosome segregation: very recently Rudd and Willard found that HACs containing *de novo* centromeres derived from either chromosome 17 or X chromosome α-satellite repeats would incur into missegregations at a higher rate than natural chromosomes, presumably the result of anaphase lag. It is an unresolved question whether this may reflect genetic or epigenetic differences *(16)*.

There is a basic problem in the study of the origins of DNA replication: apparently no hint has been found for replication to initiate at specific sites, hence the difficulty to identify consensus origin sequences. Falaschi and coworkers *(17)* identified some time ago a replication origin complex on a G-band of chromosome 19; moreover, they could identify two proteins binding in vivo to a specific sequence. A separate but related investigation was conducted to identify the enzymes, called helicases, which perform the opening of the duplex and its subsequent unwinding, thus securing the advancement of the growing replication fork *(18)*. However, the study of these putative DNA replication origins seems to reveal two patterns *(19)*: at some loci, initiation sites can be localized, as for the β-globin locus, whereas at other loci, there are apparently multiple dispersed origins, identified as initiation zones. Despite these differences, the proteins regulating replication are highly conserved from yeast to humans and models are under study, which may include a coordination of DNA replication with other chromosomal functions. In Chapter 6, Vindigni et al. describe protocols for the isolation of newly synthesized segments and the definition of the start sites of bidirectional DNA replication.

The problem of assembling HACs has been extensively discussed in recent times and is of course the main topic in this volume. We may just remember that two strategic approaches have been considered: one is the so-called "trimming down" of existing chromosomes *(20)*, which can be obtained by *in situ* fragmentation techniques; and the other may be looked at as a "bottom-up" strategy. It rests on the identification and assembly of the genetic elements required for replication, segregation, partition, and stabilization of duplex DNA molecules *(21)*.

Both approaches are presented in the volume: perhaps the most difficult task is to identify and preserve a functional centromere, without which ACs are unstable and are quickly lost. It is comforting that in both types of strategy some success can be obtained: on one hand, a human linear minichromosome, capped by two artificially seeded telomeres, has been generated *(22)*, whereas minichromosomes containing both human and mouse centromeric elements have been transmitted through the mouse germ line *(23)*; on the other hand the incorporation of large blocks of α-satellite DNA has allowed the formation of mitotically stable HACs with a functional centromere *(24)*.

De las Heras and the others of the Edinburgh group elaborate on the bottom-up approach, which, in theory, allows to build a MAC with well-defined components (*see* Chapter 10) these authors transfected a PAC vector containing human telomeric and centromeric sequences into a human cell line, obtaining in a number of cases extrachromosomal structures, which derived only from the input DNA and segregate in a stable way during cell division.

Lim and Farr, on the other hand, after reviewing the basic functions required by an engineered artificial chromosome, describe the possible manipulations of existing chromosomes, with special regard to chromosome fragmentation using cloned telomeric DNA (*see* Chapter 9) this technique has allowed the generation of minichromosomes from human X and Y chromosomes as well as neocentromere-based human minichromosomes. Another part of their work is devoted to the generation of transgenic mice carrying human extra chromosomes, an exciting advance in the study of models for human disease: in this perspective, it is encouraging that it may be also possible to induce mice to secrete and assembly human antibodies. Along this line, Kuroiwa, Tomizuka, and Ishida describe here a system based on human chromosome-derived fragments that can be used as vectors for large stretches of human DNA, thus overcoming size limitations of conventional methods (*see* Chapter 11). Moreover, these vectors can be maintained as single-copy extra chromosomes in host cells, preventing toxic overexpression or gene silencing.

A peculiar approach has been pursued by De Jongh and associates, based on the generation of satellite DNA-based ACs, also referred as artificial chromosomes expression systems (ACes), which replicate and segregate alongside the host chromosomes *(25)*. ACes possess the functional and structural sequences of natural chromosomes, including telomeres, centromeres, and replication origins. These last elements are reputed to be unknowingly distributed along the entire fragment length. Transgenic mice have been obtained by pronuclear microinjection of these artificial constructs and the examination of metaphase chromosomes from lymphocytes of manipulated mice show that ACes are maintained as discrete, independent entities and are not integrated with host chromosomes. In Chapter 12, Monteith et al. describe the procedure used to obtain these mice and discuss the implication of the relevant methodology.

Gene therapy studies using ACs are still in a very early stage of this controversial area of research, as it is for many facets of this approach. However, some interesting results have already been obtained, for instance, by Ioannou et al. *(26)*, who used a bacterially derived artificial chromosome system (BAC) to introduce targeted modifications in the host genome; however, genetic manipulation appeared difficult to control with this technique. Later on, a second-generation BAC-PAC cloning vector allowed the insertion of a 185-kb sequence containing the human β-globin gene cluster: this system seems to minimize the risk of unwanted rearrangements and allows the introduction of modifications or of reporter genes at any specific sequence *(27)*.

In Chapter 13, Orford and coworkers describe the so-called GET recombination system, which is expected to facilitate the introduction of a variety of modifications into genomic fragments in BAC-PAC clones. This approach may be used to introduce mutations or polymorphisms in cloned genomic sequences, allowing the study of the impact of these modifications in cell lines as well as in transgenic animals and hopefully leading to the discovery of drugs capable to overcome the effects of detrimental mutations.

2. Conclusion and Outlook

A legitimate question to raise after this overview of AC may concern the practical validity the scientific challenges presented by this research line. The students of this field must recognize that a crescendo in the last half century has been characterizing the shift of biological investigation into biomolecular and cellular intervention, with clinical attempts resulting in controversial if not tragic conclusions.

In the early 1960s, the assembly of all the 64 entries of the genetic code in the form of artificial mRNA allowed for its thorough understanding and acquisition of universal significance. A mere decade later, the first artificial gene was produced through a chemical-enzymatic "total synthesis," thanks mainly to the same research group, led by H. G. Khorana. ACs should have been legitimately seen as the next target.

The biomedical literature has not been particularly rich of reports concerning in particular HAC, as we have seen; but just a few very recent articles on their transfer into mammalian hosts are remarkable and must be quoted here. They should be taken as representing a strong confirmation that the field is lively, suitable for interactions and synergies with other advanced research efforts and thus likely to produce concrete achievements in a not too distant future. We have already mentioned the successful transfer of fragments of human chromosomes into mice *(28)* and, more recently, into bovines *(29)*. Of particular interest here is the fact that the selected chromosome fragments contained the megabase-long sequences harboring all the information required for the correct synthesis and processing of both the heavy and the light chains of human immunoglobulins: this laid the foundation for a large-scale production of human polyclonal antibodies. In this regard, particular attention deserves the effort aimed at cloning the transchromosomic bovines harboring a HAC loaded with the unrearranged Ig heavy (H) and light (κ) chain sequences. Even if the problems causing reproductive cloning to be plagued by too low yield (not higher than 1%) and poor health of the survivors remain mostly unresolved, this finding shows that human immunoglobulin genes undergo correct somatic rearrangements and expression in the bovine spleen cells, where peculiarly this process occurs (differently from men and mice, where it takes place in bone marrow): the antibodies are correctly synthesized, matured, secreted, and detected in the blood of a handful of healthy newborn calves. Last but not least, the findings indicate that the HACs are retained at a high rate both at mitosis and meiosis.

The secrets of this success may reside in the fact that the 10-mb HAC has been assembled through a series of in vivo manipulations, involving first telomere fragmentation and then Cre-loxP-directed translocation between two large human DNA fragments, derived from chromosomes 14 and 22, where the two genes naturally reside and where the loxP sites have been introduced by homologous recombination. Also relevant was the series of microcell-mediated chromosome transfers: from chicken recombination-proficient cells (DT20) as primary hosts where fusion of the two chromosome chunks has taken place, to Chinese hamster ovary cells for structural analysis and adaptation to a mammalian cellular environment, and finally to bovine fetal fibroblasts. For the purpose of cloning these were fused to bovine enucleated oocytes, which were routinely cultured in vitro and eventually transplanted in utero for proper development to fetal stage.

Fetal fibroblast cells were again recovered and cultured in vitro: these regenerated somatic cells proved superior nuclear donors for reproductive cloning. All or some of these innovative steps may have added efficiency to the production of transchromosomic cows and their cloning.

In conclusion, the range of options presented to the attention of the scientific community and hopefully to the safe fruition of humankind is becoming wider and tempting: relevant to this is the announcement that structurally complete and functionally unimpaired poliovirus particles can be artificially obtained in the absence of natural template, but rather using the information stored in a chemically synthesized artificial cDNA *(30)*. This may resemble the repetition of an experiment performed by Baltimore et al. more than 20 yr ago using a "natural" cDNA template, as properly remarked

(31). A more considerate evaluation of this work may emphasize the use of an "artificial" cDNA-based, single-chromosome genome: this approach may thus be seen as paving the way to other daunting enterprises. Along this line, mention is due to the recent production of "artificial gametes" achieved by soaking washed spermatozoa into DNA solution: this controversial technique, first described almost 15 yr ago, has shown a surprisingly high efficiency in the generation of transgenic pigs by artificial insemination with the genetically modified semen, and through the selection and transfer of appropriate transgenes (e.g., the hDAF, interfering with complement action) may increase the chances of xenotransplantation *(32)*.

In many researchers' dreams, if not agendas, streamlined versions of the genome of *Escherichia coli*, the true horseback of 20th century molecular genetics, or of the naturally short-sized *Mycoplasma genitalium* genome, could well be the best candidates for what may seem one of the challenges of 21st century molecular biology: the artificial synthesis and eventual manipulation of a living cell *(33)*.

References

1. Willard, H. F. (2000) Artificial chromosomes coming to life. *Science* **290,** 1308–1309.
2. Sgaramella, V. and Eridani, S. (1996). Mammalian artificial chromosomes: A review. *Cytotechnology* **21,** 253–261.
3. Brown, W. R. A., Mee, P. J., and Shen, M. H. (2000). Artificial chromosomes: Ideal vectors? *Ophtalmic Gene.* **18,** 218–223.
4. Malferrari, G., Castiglioni, B., Rocchi, M., Sgaramella, V., and Biunno, I. (2001) Partial characterization of a minichromosome derived from human chromosome 13. *Transgenics* **3,** 243–250.
5. Schlessinger, D. (1990) Yeast artificial chromosomes: tool for mapping and analysis of complex genomes. TIG **6,** 248–258.
6. Sgaramella, V., Ferretti, L., Damiani G., and Sora, S. (1990) A procedure for cloning restriction fragments of DNA as single inserts in yeast artificial chromosomes. *Biochem. Int.* **20,** 503–510.
7. Mucha, M., Lisowska, K., Goc, A., and Filipski, J. (2000) Nuclease-hypersensitive chromatin formed by a CpG island in human DNA cloned in an artificial chromosome in yeast. *J. Biol. Chem.* **275,** 1275–1278.

8. ISCNDB. (2001) International system for chromosome nomenclature of domestic bovids. Di Bernardino D. et al. (L. Jannuzzi, Coordinator). *Cytogenetics Cell Genet.* **92,** 283–299.

9. Ferguson-Smith, A. C. and Surani, M. A. (2001) Imprinting and the epigenetic asymmety between parental genomes. *Science* **293,** 1086–1093.

10. Hendrich, B. and Bickmore, W. (2001) Human diseases with underlying defects in chromatin structure and modification. *Human Molec. Genetics* **10,** 2233–2242.

11. Muller, S., Scaffidi, P., Degryse, B., et al. (2001) The double life of HMGB1 chromatin protein: architectural factor and extracellular signal. *EMBO J.* **20,** 4337–4340.

12. Riethman, H. C., Xiang, Z., Paul, S., et al. (2001) Integration of telomere sequences with the draft human genome sequence. *Nature* **409,** 948–950.

13. Corey, D. R. (2002) Telomeres inhibition, oligonucleotides and clinical trials. *Oncogene* **21,** 631–637.

14. Du Sart, D., Cancilla, M. R., Earle, E., et al. (1997) A functional neocentromere formed through activation of a latent human centromere and consisting of non-α-satellite DNA. *Nat. Genet.* **16,** 144–153.

15. Harrington, J. J., Van Bokkelen, G., Mays, R. W., et al. (1997) Formation of de novo centromeres and constitution of first-generation artificial chromosomes. *Nat. Genet.* **4,** 345–355.

16. Rudd, M. K. and Willard, H. F. (2002). Segregation of natural and artificial chromosomes. *Am. J. Hum. Genet., Suppl.* **71,** 217.

17. Falaschi, A. (2000) Eukaryotic DNA replication: A model for a fixed double replisome. *TIG* **16,** 88–92.

18. Vindigni, A., Ochem, A., Triolo, G. and Falaschi, A. (2001) Identification of human DNA helicase V with the far upstream element-binding protein. *Nucleic Acids Res.* **29,** 1061–1067.

19. Gilbert, D. M. (2001) Making sense of eukaryotic DNA replication origins. *Science* **294,** 96–100.

20. Raimondi, E., Ferretti, L., Young, B. D., Sgaramella, V., and De Carli, L. (1991). The origin of a morphologically unidentifiable human supernumerary minichromosome traced through sorting, molecular cloning and in situ hybridization. *J. Med. Genet.* **28,** 92–96.

21. Grimes, B. and Cooke, H. (1998) Engineering mammalian chromosomes. *Hum. Mol. Genet.* **7,** 1635–1640.

22. Mills, W., Critcher, R., Lee, C., and Farr, C. J. (1999) Generation of a 2–4 Mb human X centromere-based minichromosome by targeted telomere-associated chromosome fragmentation in DT40 cells. *Hum. Mol. Genet.* **8,** 751–761.

23. Shen, M. H., Yang, J., Loupart, M.-L., et al. (1997) Human minichromosomes in mouse embryonal stem cells. *Hum. Mol. Genet.* **6,** 1375–1382.

24. Kuroiwa, Y., Tomizuka, K., Shinohara, T., et al. (2000) Manipulation of human minichromosomes to carry greater than megabase-sized chromosome inserts. *Nat. Biotech.* **18,** 1086–1091.

25. De Jongh, G., Telenius, A. H., Telenius, H., et al. (1999). A mammalian artificial chromosomes pilot production facility: Large scale isolation of functional satellite DNA-based artificial chromosomes. *Cytometry* **35,** 129–133.

26. Ioannou, P., Amemiya, C.Y., et al. (1994). A new bacteriophage P1-derived vector for the propagation of large human DNA fragments. *Nat. Genet.* **6,** 84–89.

27. Narayanan, K., Williamson, R., Zhang, Y., Stewart, A. F., and Ioannou, P. (1999) Efficient and precise engineering of a 200 kb β-globin human artificial chromosome in E.coli DH10B using an inducible homologous recombination system. *Gene Ther.* **6,** 442–447.

28. Tomizuka, K., Shinohara, T., Yoshida, H., et al. (2000). Double trans-chromosomic mice: maintenance of two individual human chromosome fragments containing the Ig heavy and κ loci and expression of fully human antibodies. *Proc. Natl. Acad. Sci. USA* **97,** 722–727.

29. Kuroiwa, Y., Kasinathan, P., Choi, Y. J., et al. (2002) Cloned transchromosomic calves producing human immunoglobulin. *Nat. Biotechnol.* **20,** 889–894.

30. Cello, J., Paul, A. V., and Wimmer, E. (2002) Chemical synthesis of poliovirus cDNA: generation of infectious virus in the absence of natural template. *Science* **297,** 1016–1018.

31. Block, S. M. (2002) A not-so-cheap stunt. *Science* **297,** 769.

32. Lavitrano, M., Bacci, M. L., Forni, M., et al. (2002) Efficient production by sperm-mediated gene transfer of human decay accelerating factor (hDAF) transgenic pigs for xenotransplantation. *Proc. Natl. Acad. Sci. USA* **99,** 14,230–14,235.

33. Marshall E. (2002) Venter gets down to life's basics. *Science* **298,** 1701.

2

Methodologies Applied to Domestic Animal Chromosomes

Leopoldo Iannuzzi

1. Introduction

Chromosomes of domestic animals have attracted the attention of both scientists and breeders because chromosomal abnormalities have been strictly correlated with the reduced fertility in cattle carrying rob(1;29) *(1)*. Domestic animal cytogenetics has expanded noticeably, extending its interest not only to clinical cytogenetics but also to evolutionary and, more recently, molecular cytogenetics (gene mapping). Chromosomes of domestic animals, especially those of bovids, are very difficult to study because all autosomes of cattle, goats, and dogs, most of them from sheep and river buffalo, and many of them from horses are acrocentric with a decreasing, but similar, size.

Chromosome banding techniques have been largely applied in domestic animals. International chromosome nomenclatures have established standard banded karyotypes for cattle, sheep, goat, pig, horse, river buffalo, and rabbit *(2–7)*, although problems concern-

From: *Methods in Molecular Biology, Vol. 240:*
Mammalian Artificial Chromosomes: Methods and Protocols
Edited by: V. Sgaramella and S. Eridani © Humana Press Inc., Totowa, NJ

ing some chromosomes, especially for cattle, goat, and sheep, have only recently been solved. Indeed, only when molecular markers were assigned to each cattle and sheep chromosomes *(8)* and the same markers were applied on both Q/G- and R-banded cattle chromosome preparations *(9)* were Q-, G-, and R-banded standard karyotypes of cattle, sheep, and goat arranged using only one common chromosome nomenclature *(10)*.

This represents an important point of reference for further studies on domestic bovid chromosomes.

The recent development of molecular cytogenetics also in domestic animals offers another important tool to the cytogeneticists. The use of specific molecular markers, or of chromosome painting probes, and the fluorescence *in situ* hybridization (FISH) technique permit considerable advances in our knowledge of chromosome homologies among related and unrelated species and the straightforward identification of chromosome abnormalities (mainly reciprocal translocations and paracentric inversions) that normally escape the cytogenetic analyses, especially when acrocentric chromosomes are involved.

In this chapter, protocols for blood cell cultures, CBA-, RBA-, RBG-, and GBG-banding techniques, the *in situ* hybridization technique, and signal detection will be described for their easy use on domestic animal chromosomes.

2. Materials

1. Peripheral blood samples are collected by sterile tubes containing sodium heparin (vacutainer system).
2. Mitogen for blood lymphocyte cultures: Concanavalin A (Sigma, C-2010). Dissolve 50 mg Concanavalin A in 50 mL Puck's solution, pH 7.0, then filter with sterile 0.2-micron filter, aliquot in 5-mL sterile tubes or glass flash, and store at –20°C.
3. Physiological solution: Puck's solution 8.0 g/L NaCl, 0.4 g/L KCl, 1.0 g/L glucose, 0.35 g/L NaHCO. Bring to pH 7.0 with 1 *N* HCl.
4. Colcemid for cell cycle block at the metaphase. KaryoMax Colcemid solution (Gibco-BRL, cat. no. 15210-040).

5. BrdU (5-bromodeoxyuridine, thymidine base analog for replicating G and R banding; Sigma B-5002): Dissolve 20 mg BrdU in 20 mL Puck's solution, then filter with a 0.2-micron sterile filter and aliquot in 5-mL sterile tubes. Use 0.2 mL of this solution on 10 mL cell culture to obtain a WS at 20 µg/mL.

6. Methotrexate (MTX; Ametopterine, Sigma A-6770) for cell cycle synchronization. Dissolve 10 mg MTX in 10 mL distilled water (SS1 = 1 mg/mL), then dilute 0.5 mL SS1 in 19.5 mL Puck's solution (pH 7.0), filter with a 0.2-micron sterile filter and aliquot in 5-mL sterile tubes or glass flasks (SS2 = 25 µg/mL). Use 0.2 mL of SS2 in 10 mL cell culture to arrive at a final WS of 0.5 µg/mL. Store both SS1 and SS2 at –20°C.

7. Ethidium bromide (EB, Sigma E-8751) for more elongated G-banded chromosomes. Dissolve 20 mg EB in 20 mL distilled water (SS = 1 mg/mL), then filter with a sterile 0.2-micron filter and aliquot in 5-mL sterile tubes. Use 50 µL of SS in 10 mL cell culture to obtain a final concentration of 5 µg/mL.

8. 2X SSC (g/L). NaCl 17.53, 3-sodium citrate 8.82. Bring to pH 7.0 with 1 N HCl.

9. Phosphate buffer (P-buffer). Mix 39.0 mL solution **A** (6.95 g of NaH_2PO_4 in 250 mL distilled water) with 61.0 mL solution **B** (35.8 $gNa_2HPO_4.12H_2O$ in 500 mL distilled water).

10. Hoechst 33258 (Bisbenzimide, Sigma B-2883) for staining. Dissolve 10 mg Hoechst 33258 in 20 mL distilled water (SS = 0.5 mg/mL), aliquot in a 1-mL tube and store at –20°C until use. Dilute 1 mL of this solution in 20 mL distilled water for staining (WS = 25 µg/mL) and store at 4°C.

11. Hoechst 33258 (Bisbenzimide, Sigma B-2883) for cell cultures (R banding). Dissolve 20 mg Hoechst 33258 in 10 mL distilled water (SS = 2 mg/mL), then filter with a 0.2-micron sterile filter and aliquot in 5-mL sterile tubes. Use 0.2 mL of SS in 10 mL cell culture to reach 40 µg/mL as WS.

12. Biotin incorporation. BioNick labeling system kit (Gibco-BRL/Life technology, cat. no. 18247-015).

13. Hybridization solution (HS): 5 mL formamide (J. T. Baker, cat. no. 7042), 1 mL 20X SSC, and 2 mL dextran sulphate (Sigma, cat. no. D-8906 at 50%) = 8 mL HS. Mix the solution very well, and divide it in aliquots (1 mL each) and store at –20°C until use.

14. Bovine COT-1 DNA for *in situ* suppression of repetitive sequences present in the genomic probes (bovids) (Applied Genetic Laboratory [AGL], Inc., Melbourne, FL).
15. FISH detection kit (FITC–avidin): Oncor, Biotin-FITC kit S1333-BF.
16. FISH detection kit (anti-avidin): Oncor, same kit as for FITC-Avidin.
17. PN buffer for posthybridization washing buffer: 13.8 g/L NaH_2PO_4 (0.1 *M*), 35.8 g/L Na_2HPO_4 (0.1 *M*), Nonidet P-40 (0.1%). Bring the solution to pH 8.0 with 5 *N* NaOH.
18. Antifade (100 mL): 0.1 g 1,4-phenylendiamin (Sigma, cat. no. P-6001), PBS [NaH_2PO_4 (0.2 *M*) + Na_2HPO_4 (0.2 *M*) + NaCl (0.15 *M*)] 10 mL, glycerol (Rudi Pont, cat. no. 17500-11) 90 mL. Aliquot in 10-mL tubes and store at –20°C.
19. Antifade/Hoechst 33258 (2 µg/mL) solution: Antifade 50 mL, H-33258 0.2 mL from SS at 0.5 mg/mL (2 µg/mL, final concentration). The Oncor kit Biotin-FITC S1333 also contains antifade and antifade/propidium iodide solution.

3. Methods

3.1. Normal Cell Cultures

1. Add 0.8–1.0 mL peripheral blood sample to a 15-mL sterile tube or a 50-mL sterile flash (the same as that used for fibroblast cell cultures) containing 8.0 mL of TC medium (McCoy's 5A modified or RPMI 1640, Gibco), 1.0 mL of inactivated (at 56°C for 30 min) fetal or bovine calf serum, Concanavalin A (15 µg/mL, final concentration), penicillin/streptomycin (0.1 mL), L-glutamine (0.05 mL when present in the medium, 0.1 mL when not), and one drop of sterile sodium heparin (this prevents coagulation problems). Other mitogens, such as the Pokeweed or the PHA, can be used instead of Concanavalin A. However, the latter offers the best results as mitogen and is cheaper than Pokeweed and PHA. Only for horse and donkey cell cultures, Pokeweed mitogen must be preferred to the Concanavalin A.
2. Store cell cultures at the 37.8°C in a normal incubator or at 37.5°C in a CO_2 incubator (with CO_2 at the 4.5%). When using tubes, keep them with the highest inclination to improve cell growth.
3. Gently agitate cell cultures once a day.
4. Add 20–50 µL Colcemid (depending on species and expected chromosome contraction) 1 h before the harvesting (*see* **Note 1**).

5. Top spin at 1200*g* for 8 min, remove the supernatant, and add KCl 0.75 *M* (0.56 g %) drop by drop to arrive at 2 mL by shaking the tube gently. Mix cells thoroughly by using Pasteur pipet, and then add more solution to arrive at 14 mL. Mix cells with a Pasteur pipet and store the cell suspension at 37°C for 20 min. Then, add 1 mL of fix solution (FS) (acetic acid/methanol 1:3) and mix (*see* **Note 2**).

6. Top spin at 1000*g* for 10 min, remove the supernatant, and add (drop by drop) 2 mL FS. Then, mix thoroughly with a Pasteur pipet (be sure to break down cell clusters when present) and add more fix solution to arrive at 10 mL. Mix with a Pasteur pipet and store at room temperature for 20 min (*see* **Note 2**).

7. Top spin at 1000*g* and remove the supernatant. Add 5 mL of FS, mix with a Pasteur pipet, and store at room temperature for 10 min.

8. Repeat as in **step 7** and store at 4°C overnight.

9. Repeat as in **step 7**.

10. Repeat as in **step 7** by adding 0.5–1.0 mL fresh FS (the quantity depends on pellet size).

11. Spread two drops of cell suspension on slides previously cleaned with ethanol and immerse in cold distilled water.

12. Air-dry the slides and check cell density with a microscope by using phase-contrast.

3.2. BrdU-Treated Cell Cultures

Follow the protocol as for normal cultures with a few differences.

3.2.1. Late BrdU Incorporation (R Banding)

1. Add BrdU (20 µg/mL final concentration) and Hoechst 33258 (40 µg/mL final concentration) to cell cultures 6 h before harvesting.

2. Add 20–40 µL Colcemid 30–60 min before harvesting (*see* **Notes 1, 3, 4, 5, 6, 7**).

3.2.2. Early BrdU Incorporation (G Banding) for Cattle, River Buffalo, Horse, and Donkey (11,12)

1. Add BrdU (20 µg/mL, final concentration) and MTX (0.5 µg/mL, final concentration) to the cell cultures 20–22 h before harvesting (afternoon).

2. Top spin cell suspension at 1200*g* after 16–18 h (early morning) and eliminate the supernatant.
3. Wash cells once with 15 mL Puck's solution or with the same medium, then spin at 1200*g* for 8 min and remove the supernatant.
4. Add fresh TC medium as in normal cultures containing also thymidine (10 µg/mL, final concentration) and store at 37.5°C (normal incubator) or 37.7°C (CO_2 incubator) for 5.5 h.
5. Add 20 µL of Colcemid 30 min before harvesting (*see* **Notes 8–11**).

3.2.3. Early BrdU Incorporation (G Banding) for Sheep, Goat, Pig, Dog, Rabbit, and Chicken

1. Add BrdU (20 µg/mL, final concentration) to cell cultures 8 h before harvesting (early morning);
2. After 2.5 h top spin at 1200*g*, remove the supernatant and follow the same protocol described above (**steps 3–5**) (*see* **Notes 8–11**).

3.3. Banding Techniques

Several banding techniques are available. I will refer only to those routinely used in my laboratory because they offer high-resolution banding patterns and the protocols are successfully repeatable.

3.3.1. CBA Banding

Use slides obtained from both normal and BrdU-treated cell cultures and stored at room temperature for at least 1 wk. This protocol is a modification of the original Sumner *(13)* protocol.

1. Immerse slides in HCl 0.1 *N* for 30 min at room temperature, then wash them with distilled water and air-dry.
2. Immerse slides completely in $Ba(OH)_2$ (5% filtered solution) at 50°C for 20–30 min. We normally use two slides per animal and two different treatment times (20 and 30 min) with $Ba(OH)_2$.
3. Because the slides are covered by $Ba(OH)_2$ solution, aspirate the white coat before to removing the slides or wash the slides directly in the same Coplin jar with tap water, then with distilled water.
4. Air-dry slides at 40°C for 5 min and immerse them in 2X SSC at 60°C for 30 min and then for 15 s in 2X SSC at room temperature.

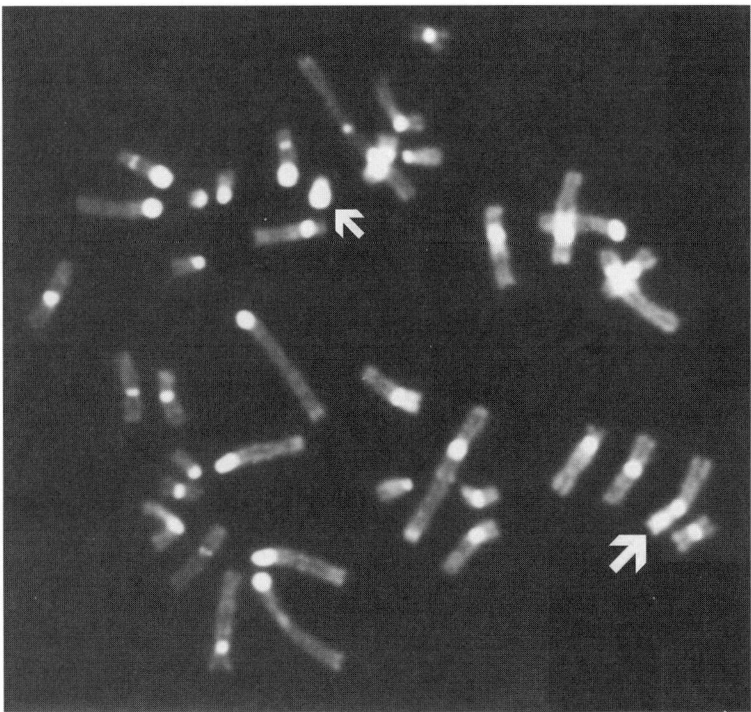

Fig. 1. CBA-banding in a male pig metaphase plate (2n = 38, XY). X (large arrow) and Y (small arrow) chromosomes are indicated. Notice the strong fluorescence (C band positive) in the entire Y chromosome.

5. Dehydrate slides in 75% and 95% alcohol series (3 min each) and air-dry.
6. Stain with acridine orange (0.1% in P buffer, pH 7.0) for 1 h. Then wash in tap and distilled water and air-dry.
7. Mount slides in P buffer with glass coverslip, press coverslip with paper to eliminate the excess of buffer, and seal with rubber cement.
8. Microscope observation a day later with appropriate filters (excitation filters at the 450–490 nm) (**Fig. 1**) (*see also* **Notes 3** and **4**).

3.3.2. RBA Banding

Stain slides obtained from late BrdU-incorporation cultures with acridine orange (0.1% in P buffer, pH 7.0) for 10 min and continue

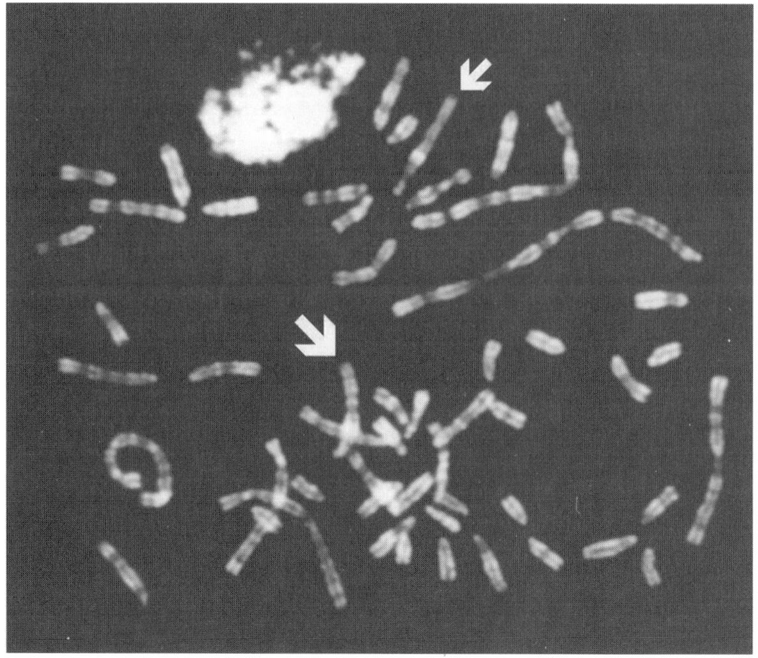

Fig. 2. RBA-banding in a female sheep early-metaphase plate (2n = 54, XX). Early (large arrow) and late (small arrow) replicating X chromosomes are indicated.

as for CBA banding (**steps 6–8**): fluorescence R banding will be performed (**Fig. 2**) (*see* **Notes 5–7**).

3.3.3. RBG Banding

1. Stain 1-wk-old (or more) slides with Hoechst 33258 (25 μg/mL in distilled water) for 20 min. Then, wash slides with distilled water and air-dry at 40°C for 10 min.
2. Mount slides with 1 mL 2X SSC (pH 7.0) using coverslip without pressure, then expose slides under UV light for 1 h at the distance of 4–5 cm from the lamp (30-W UV lamp). Wash slides with distilled water and air-dry at 40°C for 10 min.
3. Immerse slides in 2X SSC (pH 7.0) at 60°C for 30 min, then in 2X SSC at room temperature for 15 s.

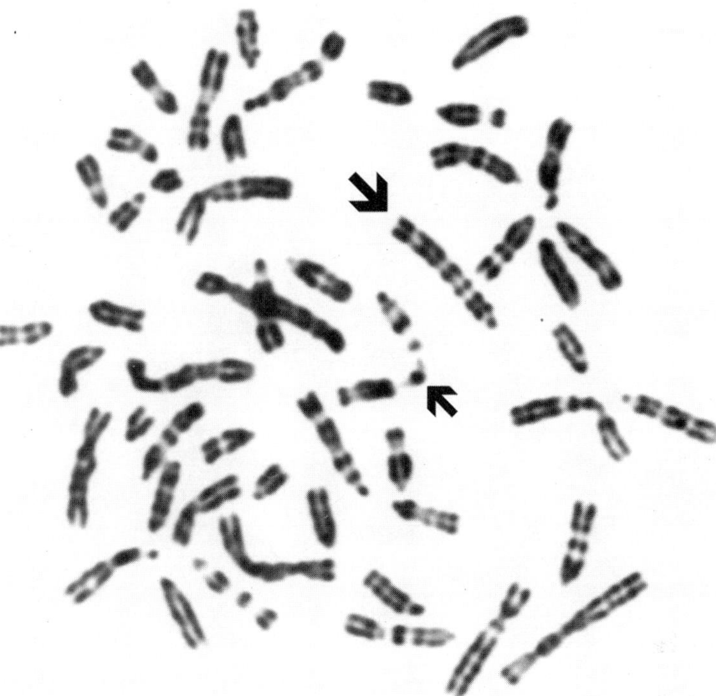

Fig. 3. RBG banding in a female river buffalo early-metaphase plate (2n = 50, XX). Early (large arrow) and late (small arrow) replicating X chromosomes are indicated.

4. Wash slides with tap and distilled water and air-dry. Then stain with Giemsa (8% in P-buffer, pH 7.0) for 30 min.
5. Microscope observation 1 d later without coverslip when slides are used for other banding techniques (C banding or Ag-NORs) or with coverslip by using Eukit as mounting slides (**Fig. 3**) *(see* **Notes 5–7**).

3.3.4. GBG Banding

Use slides from cultured treated with early BrdU incorporation and follow the same protocol used for RBG banding. Replicating G-banding patterns will be obtained (**Fig. 4**) *(see* **Notes 8–11**).

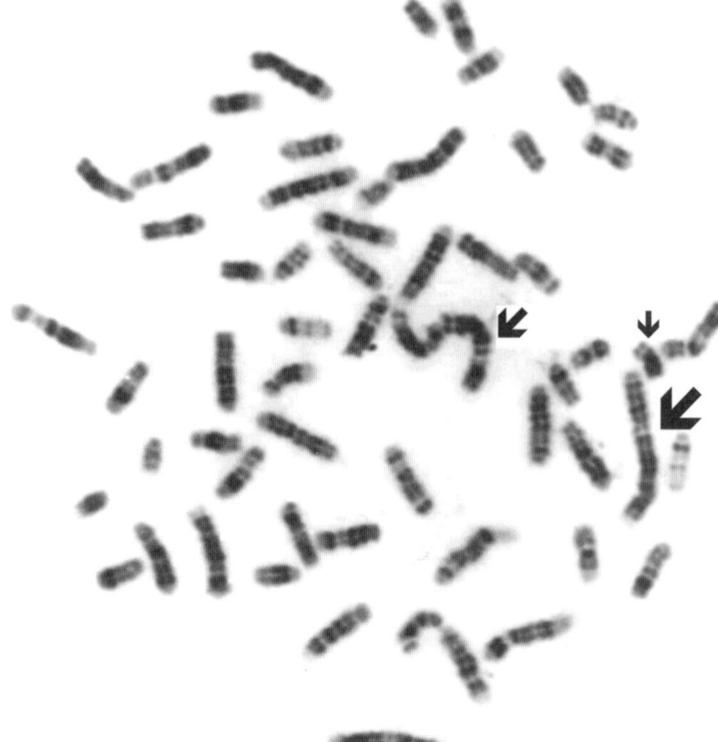

Fig. 4. GBG banding in a male cattle early metaphase plate [2n = 59, XY, rob(1;29)]. The translocated chromosome (large arrow), as well as X (medium arrow) and Y (small arrow) chromosomes are indicated.

3.4. Fluorescence In Situ Hybridization (FISH)

3.4.1. Biotin Incorporation and Probe Precipitation

Biotin-14-dATP is incorporated into 1 µg probe DNA (generally cosmids or BAC-clones) by Nick translation. Pipet the following components into a sterile 1.5-mL microcentrifuge tube on ice.

1. A quantity (µL) of probe DNA to arrive at 1 µg probe DNA, 5 µL of 10X dNTP mix, and sterile water to arrive to 45 µL and 5 µL enzyme mix (DNA Polymerase I and DNase I).

2. Close the tube, mix well, centrifuge for a few seconds at 12,000*g*, and incubate at 16°C for 1 h, then add 5 µL of stop buffer.
3. Add 100 µL of sonicated salmon sperm and then 15 µL of sodium acetate (2.5 *M*).
4. Mix well and add 300 µL cold –20°C ethanol (95%), mix well, and top spin (13,000*g*) for a few seconds.
5. Store at –20°C for at least 30 min and top spin at 13,000*g* in a cold centrifuge for 20 min.
6. Eliminate the supernatant and add 500 µL 70% cold ethanol.
7. Wash the pellet and top spin at 13,000*g* for 10 min.
8. Eliminate the supernatant and carefully air-dry the pellet (a vacuum pump system may be useful).
9. Add 33 µL hybridization solution (HB) to obtain a probe concentration at 30 ng/µL, and dissolve the pellet very well by using vortex.
10. Top spin at 13,000*g* for a few seconds and store the probe at –20°C until use (can be stored for up to 1 yr or more).

3.4.2. Probe Denaturation

1. For each *in situ* hybridization, prepare 11 µL probe DNA containing 4 µL probe stock (about 120 ng probe DNA), 6 µL HS and 1 µL bovine COT-1 DNA (for bovids only) or 1–2 µL total genomic DNA of species from which the genomic probe was prepared (*see* **Note 12**).
2. Mix with vortex, then centrifuge for a few seconds at 13,000*g*.
3. Immerse the tube containing the probe in water at 72°C for 15 min.
4. Immerse the tube containing probe DNA in water at 37.5°C for 1 h (annealing step) to suppress the repetitive sequences with bovine COT-1 DNA (or with total genomic DNA).

3.4.3. Chromosomal (Slide) Denaturation

1. Select good slides (good mitotic index and chromosome contraction) and the best slide area by recording the data (approx 2 × 2 cm) with the phase-contrast microscope.
2. Stain with Hoechst 33258 by following the same procedure as reported in **Subheading 3.3.**, **steps 1** and **2** (RBG banding).
3. Immerse slides in 70% formamide/2X SSC solution (pH 7.0) for 2.5 min at 72°C (we use two different glass Coplin jars, alternatively, to maintain the temperature at 72°C).

4. Immerse slides in an alcohol series (70%, 80%, and 96%) for 2 min each at 4°C (we keep the Coplin jar in ice) and quickly air-dry with air flush.

5. Mark the slide area (selected during **step 1**) of the *in situ* with an indelible pen (front side).

3.4.4. In Situ *Hybridization*

1. Apply the denaturated probe DNA mixture on the marked area of slides. It is preferable to achieve this step by combining probe DNA denaturation (and annealing step) with slide denaturation.

2. Mount slides with a glass coverslip (2 × 2 cm) without pressure, seal them with rubber cement and allocate slides in a moist chamber (Petri dish) and store at 37°C for 3 d (we generally hybridize during the weekend).

3.4.5. *Signal Detection and R Banding*

1. Remove coverslip and wash slides in three Coplin jars containing formamide/2X SSC (1:1), pH 7.0, at 42°C for 5 min each. Generally, the first washing is necessary to remove the coverslips only (let the coverslips be removed by themselves by gently washing the slides), and the following two washings are used to remove the nonspecific binding probe DNA.

2. Wash slides in 2X SSC, pH 7.0, at 39°C and in 2X SSC at room temperature (2 min each).

3. Wash slides in three PN buffer series (2 min each).

4. Add 20 µL FITC–avidin on the marked slide area (2 × 2 cm) and mount with plastic coverslip (2.5 × 2 cm) (pieces of parafilm can also be used) without pressure.

5. Incubate slides at 37.5°C in a moist chamber for 30 min, then repeat **step 3**.

6. Add 20 µL of anti-avidin antibody and repeat as in **step 4**.

7. Incubate slides at 37.5°C in a moist chamber for 30 min, then repeat **step 3**.

8. Add a second layer of FITC-avidin as in **step 4**.

9. Repeat **step 3** and mount slides with glass coverslips (24 × 56 cm) by using one large drop of Antifade/Hoechst 33258 solution.

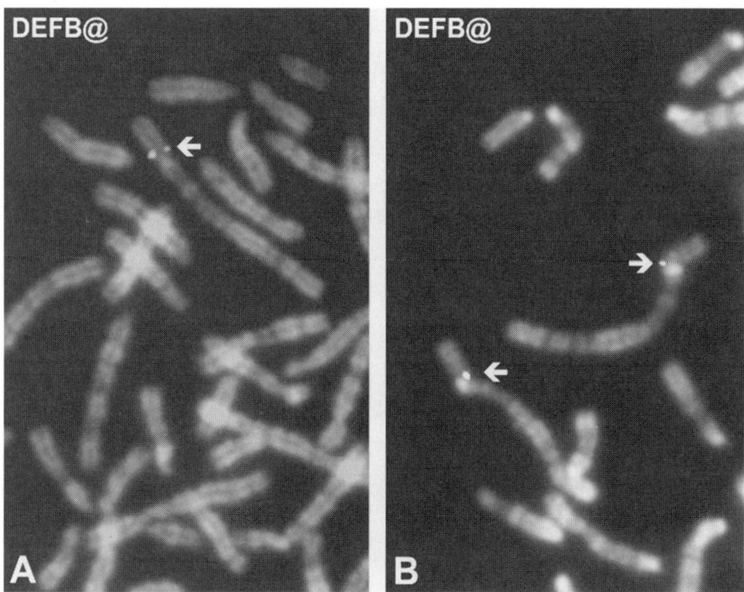

Fig. 5. Details of river buffalo prometaphase plates treated for FISH technique with a bovine BAC clone containing DEFB@. Simultaneous visualizations of the hybridization FITC signals (arrows) with RBH banding (**A**) or with RBA banding (**B**) are shown.

10. Press coverslip with paper to eliminate an excess of mounting solution.
11. Microscope observation with appropriate filter combinations with excitation filters at 340–380 nm for RBH banding (R banding by late BrdU incorporation and Hoechst 33258 staining) and at 450–490 nm for FITC signals (**Fig. 5A**).

When RBH banding is poor, slides can be counterstained with acridine orange (RBA banding) to enhance the R banding (**Fig. 5B**) by following this protocol:

1. Gently remove the coverslip and wash slides in a series of three PN buffers stored at 42°C (5 min each) and in PN buffer at room temperature (5 min), then wash slides in tap and distilled water.
2. Dehydrate slides in alcohol series (70%, 80%, and 96%) (5 min each) and air-dry.

3. Stain with acridine orange (0.1% in P buffer, pH 7.0) for 30 min, then wash slides in tap and distilled water and air-dry.
4. Mount slides in P buffer and then follow as for CBA banding (**steps 7** and **8**).
5. Microscope observation 1 d later with the same filter combination used for FITC signals (*see* **Notes 13–15**).

4. Notes

1. Colcemid treatment needs careful attention. Indeed, domestic animal chromosomes are very sensitive to this treatment. The reduction in time and quantity allow more elongated chromosome preparations to be obtained but give a lower mitotic index. Each laboratory should get the appropriate dose and time of treatment on the basis of its own needs.
2. The first fixation of cells is very important and needs special attention. When cell clusters are present, they must be broken by using a Pasteur pipet against the bottom of the tube. However, the adding of 1 mL fix solution after the hypothonic treatment not only blocks the KCl action but prevents the formation of cell clusters.
3. When observing the slides in the microscope, the C bands (CBA banding) must be very strong fluorescence, whereas the remaining part of the chromosomes must be dull fluorescence (**Fig. 1**). When other chromosome regions appear stained (fluorescent), the time of treatment in $Ba(OH)_2$ must be increased. The opposite when the C bands are very small and the chromosomes appear overtreated. When the slide treatment is good, also the nuclei show clear strong fluorescence heterochromatic regions.
4. Giemsa staining in C-banding technique (CBG banding) can be used instead of acridine orange by following the same protocol, although acridine orange is more effective than Giemsa staining. Indeed, it is possible to detect very small C bands (biarmed autosomes) or intercalar C bands, which are generally C-band negative when using Giemsa staining *(14,15)*. Furthermore, CBA-banding technique is more repeatable than CBG banding. C banding is the best banding technique to identify sex chromosomes very easily in domestic animals because their C-banding patterns differ completely from those of the autosomes *(14,15)*. Therefore, this technique is very useful to detect sex chromosome abnormalities *(16,17)*.

5. RBA banding (**Fig. 2**) offers the advantage of simplicity and its use on fresh slides. Standard RBA-banded karyotypes for cattle, sheep, goat, and river buffalo are available *(3,4,10)*.

6. RBG banding (**Fig. 3**) offers (1) higher banding pattern resolution than RBA banding (more bands compared with those achieved with RBA banding), (2) the possibility of treating slides and working on them later, and (3) keeping slides for years after staining.

7. R banding is the best banding technique to be applied routinely on domestic animal chromosomes in clinical, evolutionary, and molecular (FISH) cytogenetics *(9,16–23)*. Standard R-banded karyotypes are available for cattle, sheep, goat, river buffalo, horse, and pig *(3,4,6,7,10)*.

8. GBG banding (**Fig. 4**) is very useful when chromosomes of species must be characterized by banding techniques. Indeed, because GBG-banding patterns are exactly complementary to those obtained by R banding and are very similar to those obtained by GTG banding, this technique is a point of reference when structural G bands (GTG banding) must be compared with R-banding patterns.

9. The comparison between GBG banding with other banding techniques (GTG, RBA, and RBG banding) allowed better characterization of domestic bovid chromosome so as to obtain clear and detailed G- and R-banded ideograms following only one common banding nomenclature *(24–26)*. The only problem with this technique is that it requires early BrdU incorporation, which takes time during cell cultures, especially when BrdU is added to the cell cultures in the morning and removed later.

10. In both normal cultures and early BrdU-treated cell cultures (GBG banding), longer chromosomes can be obtained by adding ethidium bromide (5 µg/mL) 2 h before harvesting.

11. When BrdU and MTX are simultaneously added to cattle, river buffalo, horse, and donkey cell cultures to obtain GBG banding, the partial block of the cell cycle (synchronization), because of MTX, occurs during the S phase *(11,12)*. Standard GBG-banded karyotypes for cattle, sheep, goat, and river buffalo are available *(4,10)*.

12. When observing FITC signal backgrounds during FISH technique, the quantity of COT-1 DNA (or the total genomic DNA) should be increased during the *in situ* procedure (annealing step).

13. For FITC signal acquisition, the use of appropriate charge-coupled device (CCD) cameras connected with the microscope is preferred

to microphotography. Indeed, metaphases can be captured and later processed with appropriate software. Several CCD cameras are currently available. These cameras capture the images in black/white or directly in color. Generally, the former are more sensitive and pseudo-color can be given after image capturing.

14. Generally, two images of the same metaphase (with FITC signals and with RBH banding) are separately captured after the detection step and later processed by superimposing the hybridization signals on R-banded chromosomes (**Fig. 5**). Several software programs for FISH technique (image acquiring and processing) are available.

15. After the FISH-detection step, antifade/propidium iodide can be used, instead of antifade/Hoechst 33258, to obtain FITC signals against the red chromatids (propidium iodide). In R-banded preparations, the use of antifade/propidium iodide allows FITC signals to be obtained with the filter at 450–490 nm (excitation) and RBPI banding with excitation filters at 515–560 nm. However, modification of this procedure (antifade/propidium iodide at pH 11) allows simultaneous visualization of FITC signals and R-banding patterns with the combination filters at the 450–490 nm (excitation) *(9,21)*.

16. The FISH technique is a powerful tool to physically map specific loci on single chromosome bands. When two or more loci are clustered in the same bands, dual or multicolor FISH may be used *(27)*.

17. Comparative FISH mapping among species may reveal small autosomal mutations, such as that found between Bovinae chromosome 9 and Caprinae chromosome 14 *(28)*. A greater understanding may also be gained of chromosome evolution among related species by following the gene order within homologous chromosomes *(23,29,30)*, as well as comparing unrelated genomes such as those of human and bovids by establishing the chromosomal rearrangements that differentiated bovids from primates *(27,28,31–33)*. The use of specific molecular markers is also a potential tool to easily identify chromosomes involved in both numerical *(34)* and structural chromosome abnormalities *(35–37)*.

5. Conclusions

Domestic animal cytogenetics should receive special attention, especially now that several molecular markers are available and can be used in both clinical and evolutionary cytogenetics. However, more work must be done to prepare chromosome-specific painting probes and make them commercially available as for humans. This will facilitate comparisons among unrelated species (Zoo-FISH) and will accelerate the genetic improvement in domestic species because chromosomal abnormalities, especially reciprocal translocations and paracentric inversion, can easily be identified and eliminated from the animal populations. Sound collaboration among breeders, veterinary practitioners, and cytogeneticists is also essential if our domestic animal populations are to be genetically improved.

References

1. Gustavsson, I. (1969) Cytogenetics, distribution and phenotypic effects of a translocation in Swedish cattle. *Hereditas* **63,** 68–169.
2. Ford, C. E., Pollock, D. L., and Gustavsson, I., eds. Reading Conference. (1980) Proceedings of the first international conference for the standardisation of banded karyotypes of domestic animals. *Hereditas* **92,** 145–162.
3. Di Berardino, D., Hayes, H., Fries, R., and Long, S., eds. ISCNDA89. (1990) International System for Cytogenetic Nomenclature of Domestic Animals. *Cytogenet. Cell Genet.* **53,** 65–79.
4. CSKBB. (1994) Committee for standardized karyotype of river buffalo (Bubalus bubalis, 2n=50). *Cytogenet. Cell Genet.* **67,** 102–113.
5. Switonski, M., Reimann, N., Bosma, A. A., Long, S., Bartnitzke, S., Pinkowska, A., et al. (1996) Report on the progress of standardization of the G-banded canine (Canis familiaris) karyotype. *Chrom. Res.* **4,** 306–309.
6. Bowling, A. T., Breen, M., Chowdhary, B. P., et al. (Committee). ISCNH. (1997) International system for cytogenetic nomenclature of the domestic horse. *Chrom. Res.* **5,** 433–443.
7. Gustavsson, I., coordinator. (1988) Committee for the standardized karyotype of the domestic pig. Standard karyotype of the domestic pig. *Hereditas* **109,** 151–157.
8. Popescu, C. P., Long, S., Riggs, P., Womack, J., Schmutz, S., Fries, R., and Gallagher, D., eds. Texas Nomenclature. (1996) Standardization of cattle karyotype nomenclature: report of the committee for the standardization of the cattle karyotype. *Cytogenet. Cell Genet.* **74,** 259–261.

9. Hayes, H., Di Meo, G. P., Gautier, M., Laurent, P., Eggen, A., and Iannuzzi, L. (2000) Localization by FISH of the 31 Texas nomenclature type I markers to both Q- and R-banded bovine chromosomes. *Cytogenet. Cell Genet.* **90,** 315–320.

10. Di Berardino, D., Di Meo, G. P., Gallagher, D. S. Hayes, H., and Iannuzzi, L., eds. ISCNDB2000. (2001) International System for Chromosome Nomenclature of Domestic Bovids. *Cytogenet. Cell Genet.* **92,** 283–299.

11. Iannuzzi, L., Di Berardino, D., Ferrara, L., Gustavsson, I., Di Meo, G. P., and Lioi, M. B. (1985) Fluorescent G and C-bands in mammalian chromosomes by using early BrdU incorporation simultaneous to methotrexate treatment. *Hereditas* **103,** 153–160.

12. Iannuzzi, L., Gustavsson, I., Di Meo, G. P., and Ferrara, L. (1989) High resolution studies on late replicating segments (G+C-bands) in mammalian chromosomes. *Hereditas* **110,** 43–50.

13. Sumner, A. T. (1972) A simple technique for demonstrating centromeric heterochromatin. *Exp. Cell Res.* **75,** 304–306.

14. Di Meo, G. P., Perucatti, A., Ferrara, L., Palazzo, M., Matassino, D., and Iannuzzi, L. (1998) Constitutive heterochromatin distribution in pig (Sus scrofa) chromosomes. *Caryologia* **51,** 65–72.

15. Di Meo, G. P., Perucatti, A., Iannuzzi, L., Rangel-Figueiredo, M. T., and Ferrara, L. (1995) Constitutive heterochromatin polymorphism in river buffalo chromosomes. *Caryologia* **48,** 137–145.

16. Iannuzzi, L., Di Meo, G. P., Perucatti, A., and Zicarelli, L. (2000) A case of sex chromosome monosomy (2n=49,X) in the river buffalo (Bubalus bubalis). *Vet. Rec.* **147,** 690–691.

17. Iannuzzi, L., Di Meo, G. P., Perucatti, A., Di Palo, R., and Zicarelli, L. (2001) 50, XY gonadal dysgenesis (Swier's syndrome) in a female river buffalo (Bubalus bubalus). *Vet. Rec.* **148,** 634–635.

18. Hayes, H., Petit, E., and Dutrillaux, B. (1991) Comparison of RBG-banded karyotypes of cattle, sheep and goats. *Cytogenet. Cell Genet.* **57,** 51–55.

19. Iannuzzi, L. and Di Meo, G. P. (1995) Chromosomal evolution in bovids: A comparison of cattle, sheep and goat G- and R-banded chromosomes and cytogenetic divergences among cattle, goat and river buffalo sex chromosomes. *Chrom. Res.* **3,** 291–299.

20. Gustavsson, I., Switonski, M., Iannuzzi, L., Ploen, L., and Larsson, K. (1989) Banding studies and synaptonemal complex analysis of an X-autosome translocation in the domestic pig. *Cytogenet. Cell Genet.* **50,** 188–194.

21. Hayes, H., Petit, E., Bouniol, C., and Popescu, P. (1993) Localization of the alpha S2-casein gene (CASAS2) to the homologous cattle, sheep and goat chromosomes 4 by in situ hybridization. *Cytogenet. Cell Genet.* **64,** 281–285.

22. Iannuzzi, L., Gallagher, D. S., Ryan, A. M., Di Meo, G. P., and Womack, J. E. (1993) Chromosomal localization of omega and trophoblast interferon genes in cattle and river buffalo by sequential R-banding and fluorescent in situ hybridization. *Cytogenet. Cell Genet.* **62,** 224–227.

23. Iannuzzi, L., Di Meo, G.P., Perucatti, A., Incarnato, D., Schibler, L., and Cribiu, E. P. (2000) Comparative FISH-mapping of bovid X chromosomes reveals homologies and divergences between the subfamilies Bovinae and Caprinae. *Cytogenet. Cell Genet.* **89,** 171–176.

24. Iannuzzi, L. (1996) G- and R-banded prometaphase karyotypes in cattle (Bos taurus L.). *Chrom. Res.* **4,** 448–456.

25. Iannuzzi, L., Di Meo, G. P., and Perucatti, A. (1996) G- and R-banded prometaphase karyotypes in goat. *Caryologia* **49,** 267–277.

26. Iannuzzi, L., Di Meo, G. P., Perucatti, A., and Ferrara L. (1995) G- and R-banding comparison of sheep (Ovis aries L.) chromosomes. *Cytogenet. Cell Genet.* **68,** 85–90.

27. Gallagher, D. S., Schlapfer, J., Burzlaff, J. D., Womack, J. E., Stelly, D. M., Davis, S. K., et al. (1999) Cytogenetic alignment of the bovine chromosome 13 genome map by fluorescence in-situ hybridization of human chromosome 10 and 20 comparative markers. *Chrom. Res.* **7,** 115–119.

28. Iannuzzi, L., Di Meo, G. P., Perucatti, A., Schibler, L., Incarnato, D., and Cribiu, E. P. (2001) Comparative FISH-mapping in river buffalo and sheep chromosomes: assignment of forty autosomal type I loci from sixteen human chromosomes. *Cytogenet. Cell Genet.* **94,** 43–48.

29. Robinson, T. J., Harrison, W. R., Ponce de Leon, F. A., Davis, S. K., and Elder, F. F. B. (1998) A molecular cytogenetic analysis of the X chromosome repatterning in the Bovidae: Transpositions, inversions, and phylogenetic inference. *Cytogenet. Cell Genet.* **80,** 179–184.

30. Piumi, F., Schibler, L., Vaiman, D., Oustry, A., and Cribiu, E. P. (1998) Comparative cytogenetic mapping reveals chromosome rearrangements between the X chromosomes of two closely related mammalian species (cattle and goats). *Cytogenet. Cell Genet.* **81,** 36–41.

31. Iannuzzi, L., Di Meo, G. P., Perucatti, A., Schibler, L., Incarnato, D., Ferrara, L., et al. (2000) Sixteen type I loci from six chromosomes were comparatively fluorescent in-situ mapped to river buffalo

(Bubalus bubalis) and sheep (Ovis aries) chromosomes. *Chrom. Res.* **8,** 447–450.

32. Di Meo, G. P., Perucatti, A., Schibler, L., Incarnato, D., Ferrara, L., Cribiu, E. P., and Iannuzzi, L. (2000) Thirteen type I loci from HSA4q, HSA6p, HSA7q and HSA12q were comparatively FISH-mapped in four river buffalo and sheep chromosomes. *Cytogenet. Cell Genet.* **90,** 102–105.

33. Schibler, L., Vaiman, D., Oustry, A., Giraud-Delville, C., and Cribiu, E. P. (1998) Comparative gene mapping: a fine-scale survey of chromosome rearrangements between ruminants and humans. *Genome Res.* **8,** 901–915.

34. Iannuzzi, L., Di Meo, G. P., Leifsson, P. S., Eggen, A., and Christensen, K. (2001) A case of trisomy 28 in cattle revealed by both banding and FISH-mapping techniques. *Hereditas* **134,** 147–151.

35. Iannuzzi, L., Di Meo, G. P., Perucatti, A., Incarnato, D., Molteni, L., De Giovenni, A., et al. (2001) A new balanced autosomal reciprocal translocation in cattle revealed by banding techniques and HSA-painting probes. *Cytogenet. Cell Genet.* **94,** 225–228.

36. Iannuzzi, L., Di Meo, G. P., Perucatti, A., Eggen, A., Incarnato, D., Sarubbi, F., and Cribiu, E. P. (2001) A pericentric inversion in cattle Y-chromosome. *Cytogenet. Cell Genet.* **94,** 202–205.

37. Iannuzzi, L., Molteni, L., Di Meo, G. P., De Giovanni, A., Perucatti, A., Succi, G., et al. (2002) A case of azoospermia in a cattle bull carrying an Y-autosome reciprocal translocation. *Cytogenet. Cell Genet.* **95,** 225–227.

3

Mapping of the Hotspots of Recombination in Human DNA Cloned as Yeast Artificial Chromosomes

Grzegorz Ira, Jacek Król, and Jan Filipski

1. Introduction

Whole genome sequencing of several species, including our own, constitutes a good starting point for the study of various functions of the chromosome on a whole genome scale. Mapping of meiotic recombination hotspots (the sites in which recombination occurs more frequently than at an average site) in the genomes of higher organisms is one of the tasks that can now be undertaken. Hotspot mapping, in the case of the human genome, could be important for understanding the origin of some genetic diseases and cancers and might shed light on the mechanisms of evolution of our species at the molecular level.

Direct mapping of hotspots involves a time-consuming determination of the polymorphism of genetic markers in populations. Here,

From: *Methods in Molecular Biology, Vol. 240:*
Mammalian Artificial Chromosomes: Methods and Protocols
Edited by: V. Sgaramella and S. Eridani © Humana Press Inc., Totowa, NJ

we present a relatively simple way of mapping the recombination hotspots in human DNA cloned as yeast artificial chromosomes (YACs) based on the techniques developed for the most part by G. Simchen and collaborators *(1–3)*. The mapping consists of localizing meiotic double-strand breaks (DSBs), the hallmarks of recombination hotspots, in the YACs. Several lines of evidence suggest that sites of meiotic breaks mapped in yeast might be coincident with the hotspots of recombination in the human genome. First, it has been shown that the level of meiotic DSBs in YACs carrying segments of human chromosomal DNA correlates with the level of recombination of these segments in human germline cells *(1,2)*. The human chromosomal segment corresponding to a contracted region in the genetic map (recombination cold spot) shows a low level of meiotic DSBs in a YAC, whereas a YAC carrying DNA extracted from an expanded region of the map breaks frequently during meiosis in yeast cells. Second, we have found that this correlation observed in the case of the long DNA regions also holds for some specific human hotspots *(4,5)*. One such site has been localized in the YAC A85D10-carrying human β globin gene cluster, 30 kb 3' from the β globin gene. It contains a long palindrome that is surrounded by sequences showing polymorphism, which is typical for recombination hotspots. This site corresponds to the chromosomal breakpoint responsible for several independent cases of human thalassemia *(6)*. Another meiotic DSB that has been localized between β and δ globin genes in a YAC also corresponds to the recombination hotspot in human germline cells *(7)*. It has been demonstrated using the tetrad dissection analysis that this latter chromosomal segment also engages in meiotic recombination in *Saccharomyces cerevisiae (8)*. Apparently, because the meiotic chromosome structure and composition in *S. cerevisiae* follow the plan common for most eukaryotes *(9)*, the segments of DNA that are recombination-prone in human cells are also recombination-prone when cloned in yeast. This is also at least partially true in the case of mitotic chromosome. The YACs derived from regions of known chromosomal instability are very unstable in yeast. It has been shown, for example, that the YAC clone carrying the DNA extracted from the most unstable chromosomal region in the mouse genome is also very unstable on X-ray exposure of the yeast cells *(10)*. Such sequences as microsatellites, minisatellites, and palindromic repeats are under some conditions recombinogenic in both yeast and mammalian cells *(11–15)*.

Another important assumption upon which the presented technique is based is that the hotspots of recombination are coincident with the sites of DSBs appearing in the DNA at the onset of meiosis. The meiotic breaks have been most thoroughly studied in yeast but it is probably true for most of the eukaryotic organisms *(16–18)*. A recombination hotspot in yeast usually coincides with a 100–500-nucleotide long chromosomal segment in which DSBs can occur at a number of sites (different sites of breaks in different cells) *(19,20)*. The repair of these breaks results in a gene conversion, frequently accompanied by a crossover. Mapping of these sites is performed using mutant yeast strains that generate a normal level of breaks but that are unable to process them further. The accumulation of chromosomal breaks facilitates their mapping. The yeast strains most commonly used for this purpose carry the mutation *rad50S* (*see* **Note 1**) *(21)*.

Having these considerations in mind, we conclude that the sites mapped in YACs should be considered as so-called candidate hotspots in human DNA. As an example of the presented technique, we show here the localization of the hotspots of recombination in the YAC 745D12 containing the DNA segment extracted from human chromosome 6, which contains the major histocompatibility complex that encodes several highly polymorphic leukocyte antigens class 1. The meiotic DSB mapping was performed here at the level of resolution obtained with the pulse-field gel electrophoresis (5–10 kb). This allows one to localize the hotspots of recombination in a region of several hundred kilobases in a single experiment depending on the YAC length. It may be performed, if necessary, with a higher resolution using classic agarose gel electrophoresis.

The mapping of hotspots of recombination described here consists of three steps. The first step, that is, preparation of agarose plugs containing high molecular weight DNA, mapping of restriction sites and CpG islands in the YAC, and verification of the YAC integrity, is described in **Subheadings 3.1.1.** and **3.1.2.** and in

Table 1
Verification of the Integrity of the YAC 745D12

Position in the reference sequence (nt)		Mapped distance from the telomere C (kb)	Difference (kb)
*Mlu*I			
29375		95	65
66251		131	64
113317		178	64
160944		225	64
*Sfi*I			
54709		119	64
149627, 150482		214	64
*Bst*UI			
8322–10278	(4 sites, *IkBL*)	75	66
21036–22024	(4 sites, *BAT 1*)	85	64
28800–29376	(4 sites, *BAT 1*)	95	66
65019–66285	(10 sites, *MIC B*)	131	65
69080–70068	(10 sites, *MIC B*)	135	65
160066–160978	(11 sites, *MIC A*)	225	65
163442–164636	(7 sites, *MIC A*)	225	61

Comparison of the positions of the restriction sites localized within the nucleotide sequence AB000882 + AP000507, with the experimentally determined distances of these sites from the telomere C of the YAC 745D12. The multiple *Bst*UI sites correspond to the CpG islands accompanying the listed genes.

Table 1. The second step consists of preparing a diploid *rad50S* mutant yeast strain carrying the studied YAC using the strains AB1380 [YAC 745D12]; *kar1*; and *rad50S* strains. Detailed protocol of this part of the procedure has been described in vol. 54 of this series (*see* **ref. 3**) The third step, the mapping of the meiotic DSB is described in **Subheading 3.2.** and the result of the mapping is shown in **Fig. 4**.

2. Materials

2.1. Yeast Strains

The following yeast strains were used in this study.

1. AB1380 *(MATa, ura3-5, trp1, ade2-1, can1-100, lys2-1, his5, Ile⁻,* *Thr⁻)* [YAC 745D12 *(URA, TRP)*] purchased from Centre d'Etudes du Polymorphisme Humain in Paris.
2. 2474 *(MATα, ura3-52, lys2-101, ade2-101, his3Δ200, trp1Δ1, leu2Δ1, cyhʳ, kar1Δ15)*.
3. 2850 *(MATa, leu2, lys2, trp1, ura3, rad50s::ura3, can1)*.
4. YD69 *(MATα, lys2, his4, arg4, trp1, ura3, rad50s::URA3)*.

The three last strains were the kind gift of G. Simchen and D. Zenvirth from the Hebrew University of Jerusalem.

2.2. Culture Media

The following media components were purchased from Difco Laboratories.

1. Synthetic dextrose (SD), standard minimal medium: 0.17% YNB-AA/AS, 0.5% $(NH_4)_2SO_4$, 2% dextrose, containing appropriate nutrients and drugs according to auxotrophies of the strain. For a solid medium, 2% agar is added.
2. YPD (standard rich medium): 1% yeast extract, 2% peptone, 2% dextrose. For a solid medium, 2% agar is added.
3. YPA (presporulation medium): 1% yeast extract, 2% peptone, 1% potassium acetate.
4. SPM (sporulation medium): 0.1% yeast extract and 1% potassium acetate, supplemented with appropriate nutrients according to auxotrophies in the strain.
5. Nutrients and drug solutions (Sigma, St. Louis, MO; final concentrations in the SD or SPM media): L-arginine HCl (20 µg/mL), L-histidine 20 µg/mL, L-isoleucine (30 µg/mL), L-leucine (30 µg/mL), L-lysine HCl (30 µg/mL),L-threonine (200 µg/mL), L-tryptofane (20 µg/mL), adenine (20 µg/mL), uracil (20 µg/mL), cycloheximide (3 µg/mL), canavanine (60 µg/mL).

                                                               Ira et al.

2.3. Buffers, Enzymes, and Other Supplies

The following chemicals were purchased from Sigma (St. Louis, MO) if not indicated otherwise.

1. Z1: 50 mM Tris-HCl, pH 8.0; 10 mM MgCl$_2$; 1 M sorbitol; 30 mM dithiothreitol (DTT).
2. Z2: 50 mM Tris-HCl, pH 8.0; 10 mM MgCl$_2$; 1 M sorbitol; 1 mM DTT.
3. KPSS: 40 mM KH$_2$PO$_4$, pH 6.4; 10 mM MgCl$_2$; 5 mM ethylenebis-(oxyethylenenitrilo)tetra-acetic acid; 0.5 mM ethylenediamine tetra-acetic acid (EDTA); 0.2 mM DTT; 0.15 mM spermine; 0.5 mM spermidine; and 0.2 mM phenylmethyl sulfonyl fluoride (PMSF; add just before the buffer is ready to be used).
4. Lysing buffer: 1 mg/mL proteinase K (Boehringer, Mannheim), 1% SDS, 250 mM EDTA.
5. Stocking buffer: 10 mM Tris-HCl, pH 8; 50 mM EDTA; 1 M NaCl.
6. 10X TBE buffer: 890 mM Tris, 890 mM boric acid, 20 mM EDTA, pH 8 (used diluted 1:20 for electrophoresis).
7. 200 mM PMSF dissolved in isopropyl alcohol (stored –20°C).
8. 10 mg/mL EtBr.
9. 500 mM EDTA, pH 8.0.
10. 2% Low-melting point agarose (FMC BioProducts, Rockland, ME) prepared on KPSS buffer w/o spermine, spermidine, and PMSF.
11. 1% Agarose (FMC BioProducts, Rockland, ME) prepared on 0.5X TBE buffer.
12. YAC telomere-specific probe (*see* **Note 2**).
13. Glucanex® (crude preparation of β-glucanase, Novo Nordisk Ferment, Ltd): solution 200 mg/mL H$_2$O (stored –20°C).
14. *Sfi*I, *Mlu*I, and *Bst*UI restriction enzymes and the appropriate restriction buffers as suggested by the manufacturer (New England BioLabs).
15. Lambda ladder (molecular weight standard, New England BioLabs).
16. Reagents for Southern transfer and hybridization.
17. Reagents for random primed DNA labeling, including ^{32}P-labeled α-dATP (Amersham, Pharmacia Biotech).
18. Nytran® superchange nylon membrane (Schleicher & Schuell).

2.4. Special Equipment

Pulse-field gel electrophoresis apparatus (CHEF-DR® II, Bio-Rad).

3. Methods

The mapping of both meiotic DSBs and restriction sites was performed here by indirect end-labeling. This technique is commonly used to determine the distance between the breaks generated in a DNA molecule and one of its extremities. Such extremities are most frequently introduced by restriction enzyme digestion (**Fig. 1**). In the experiments described below, the sites of breakage were localized using the ends of the YAC as a reference point. To that purpose, the mixture of the molecules resulting from a partial digestion of the YAC was fractionated by agarose gel electrophoresis and transferred to a positively charged nylon membrane. The hybridization of the membrane with a ^{32}P-labeled probe, specific to one end of the YAC, revealed the unbroken YAC DNA molecules as well as their products of breakage. As a probe, we used a fragment of pBR322 (*see* **Note 2**).

3.1. Checking the YAC Integrity

Because the YAC libraries contain a large proportion of chimeric clones *(22)*, it is strongly recommended to verify the integrity of the selected artificial chromosome by restriction site mapping and comparison of the obtained map with the reference. The enzymes most frequently used to that purpose are the so-called rare cutters like *Not*I, *Sfi*I, and so on. Another class of useful enzymes is the isoschisomers that recognize the CGCG tetranucleotide *(23)*. The distribution of the latter in the human chromosomes coincides for the most part with that of CpG islands, the GC- and CpG-rich regions associated with many human genes. Mapping of these islands helps to localize the genes in the studied YAC. In addition, given that the CGCG tetranucleotides are clustered, such a map is less sensitive to allelic variations due to the differences in their number and position. The chromosomal region containing the hotspot mapped here has been independently sequenced by two groups *(24,25)* who used two sets of cosmid clones derived from two different sets of YACs. The map (**Fig. 2**) of YAC 745D12, which has been used by Shiina et al. *(25)* and also in this study, is based on the

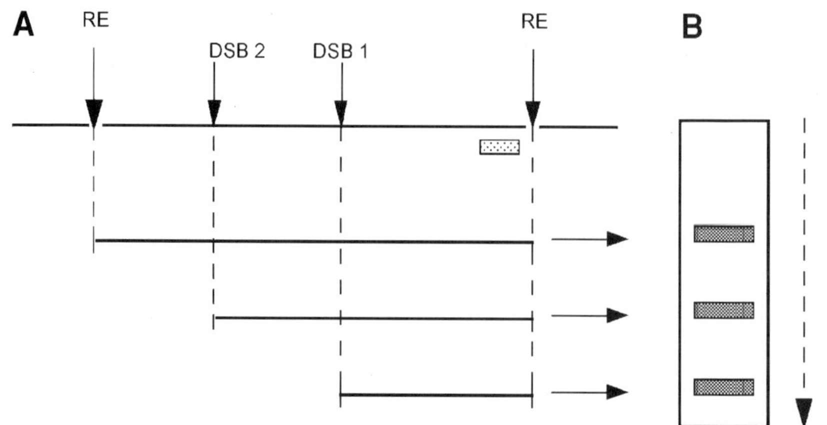

Fig. 1. Mapping of DNA breakage sites by indirect end-labeling. (**A**) A chromosome that contained breakage site DSB 1 and DSB 2 was terminally digested by a restriction enzyme recognizing the sites Re. The rectangle indicates the position of the sequence recognized by the probe. (**B**) After the electrophoresis (direction of which is shown by the broken vertical arrow) and autoradiography with the labeled probe, the positions of the sites 1 and 2, with respect to one end of the restriction segment, are calculated from the distances of migration of the DNA fragments.

published reference sequences (accession numbers AB000882 and AP000507, http://www.ncbi.nlm.nih.gov:80/entrez/). The studied region contains a putative hotspot of recombination in humans located between the genes *MIC A* and *MIC B*, as determined by allele-frequency studies *(26)*. Three restriction enzymes were chosen to check the integrity of the YAC. Two of them are rare cutters, *Mlu*I (ACGCGT) and *Sfi*I (GGCCN$_4$NCCGG), and the third one, the *Bst*UI (CGCG), is the CpG-island diagnostic enzyme. It cuts several times within each of the CpG islands accompanying these genes. The positions of the restriction sites in the reference sequences were determined using the WebCutter program (http://www.firstmarket. com/cutter/) and have been compared (**Table 1**) with the positions of the recognition sites localized experimentally in the YAC by indirect end-labeling. The results of mapping of the *Mlu*I and *Bst*UI sites are presented in the **Fig. 3**.

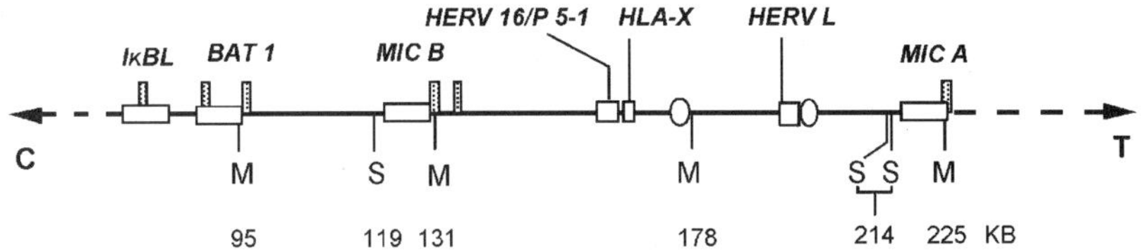

Fig. 2. Map of YAC 745D12. Genes and endogenous retroviruses are represented by white rectangles, the CpG islands by dark vertical rectangles, and the repetitive sequences by ovals. The positions of restriction sites *Mlu*I (M) and *Sfi*I (S) were determined by indirect end-labeling of the YAC (**Table 1**) and are indicated by the numbers below the letters. C and T indicate telomeres proximal and distal to the centromere respectively. The positions of principal landmarks were taken from **refs. *24*** and ***25***.

3.1.1. Preparation of Agarose Plugs Containing Chromosomal-Size DNA

1. Use one colony of the yeast strain AB1380 containing YAC 745D12 to inoculate the SD medium supplemented with adenine, lysine, histidine, isoleucine, and threonine and grow to OD_{600} of 1.
2. Inoculate 200 mL YPD medium with 100 µL preculture. Grow the yeast overnight at 30°C with good aeration up to OD_{600} 0.7 to 1.0.
3. Spin the cells at 1200*g* for 5 min. Resuspend the pellet in 10 mL buffer Z1 and incubate for 20 min at room temperature.
4. Collect the cells and wash in 10 mL buffer Z2, centrifuge as before.
5. Suspend the pellet in 10 mL buffer Z2, add 200 µL Glucanex® solution, and incubate 45–60 min at 30°C (*see* **Note 3**).
6. Spin the spheroplasts at 500*g*, resuspend the pellet in 3 mL buffer Z2, and transfer the suspension into four 2-mL Eppendorf tubes. Centrifuge 1 min in Microfuge at 1000*g*.
7. Suspend each pellet in 1 mL ice-cold KPSS buffer and mix gently (*see* **Note 4**). Spin for 15 s in the microfuge.
8. Resuspend the pellets each in 1 mL ice-cold KPSS buffer and spin for 15 s in microfuge. Repeat the washing three times. Depending on the starting cell density, the volume of the last wash might be reduced and the last centrifugation might be performed in two Eppendorf tubes.

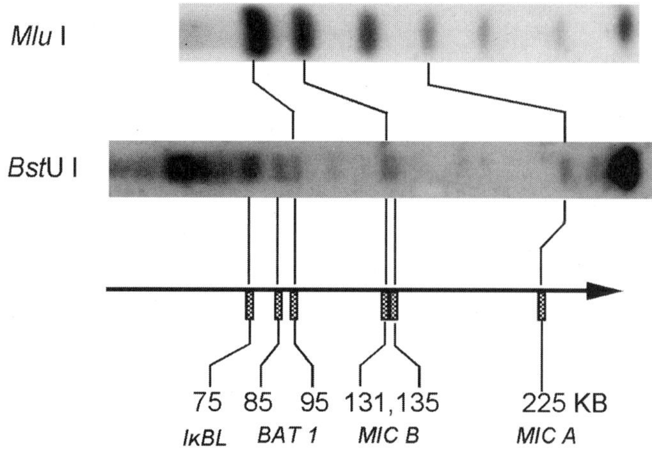

Fig. 3. Mapping the CpG islands in YAC 745D12. The DNA embedded in agarose blocks has been partially digested by restriction enzymes *Mlu*I and *BstU*I, the last having multiple recognition sites within CpG islands. The resulting products of partial digestion were separated by pulse-field gel electrophoresis, transferred to the positively charged Hybond nylon membrane, hybridized to radioactively labeled probe recognizing the telomere C and are visualized as dark bands in the X-ray sensitive photographic film (*Mlu*I lane and *BstU*I lane). The clusters of CpG dinucleotides corresponding to CpG islands accompanying the genes located in the studied regions are represented by dark rectangles in the map of the YAC. Three out of the four *Mlu*I recognition sites in this region are also located in the CpG islands (**Table 1**).

9. Measure the approximate volume of the pellets, suspend each pellet in an equal volume of KPSS buffer at the room temperature, and mix the suspension with the same volume of warm (50°C) 2% low-melting point agarose prepared in KPSS without spermine, spermidine, and PMSF (final concentration of agarose will be 1%).

10. Mix the agarose thoroughly with the suspension and quickly pour into molds (Bio-Rad).

11. Let the agarose solidify on ice and transfer the plugs to the lysing buffer (proK-SDS). Incubate overnight at 50°C.

12. Plugs prepared in this way will contain about 2 µg DNA each and can be stored for up to 6 mo in the stocking buffer at 4°C.

3.1.2. Mapping the Restriction Sites in YACs

1. Wash 10 agarose plugs containing YAC DNA with at least five changes of 10 mL appropriate 1X restriction enzyme buffer, with gentle shaking, in the cold room. The first wash should contain 0.2 mM PMSF, in addition to 1X restriction enzyme buffer, to inhibit the proteinase K.
2. Place the agarose plugs (in duplicate, to have a back-up plug if one of them is damaged during manipulation) in 2-mL Eppendorf tubes in 200 µL restriction buffer.
3. Add serial dilutions of an appropriate restriction enzyme (0.05, 0.1, 0.5, 1.0, and 5.0 U of the restriction enzyme per tube). Digest overnight at the standard conditions (*see* **Note 5**).
4. Stop the reaction by adding 20 µL 0.5 M EDTA.
5. Wash the plugs briefly in 0.5X TBE buffer.
6. Put the plugs into the slots of the slab of 1% agarose prepared in 0.5X TBE buffer with a spatula. Gently press the plugs against the front wall of the slot.
7. Place the plugs containing undigested control and molecular weight standard (Lambda ladder).
8. Seal the plugs with warm (35°C) LMP agarose. Let solidify in the refrigerator.
9. Put the slab into the electrophoresis chamber containing fresh 0.5X TBE buffer, start the cooler and the pump, and let the system run for 20 min before starting electrophoresis.
10. Electrophores samples according to manufacturer instructions.
11. Stain gel with EtBr, photograph the gel, and transfer the fractionated DNA to positively charged membrane by alkaline Southern transfer.
12. Hybridize the membrane with radioactively labeled probe C (*see* **Note 2**). Wash, expose to the X-ray film, and develop the film, or expose the film in a Phosphoimager cassette. Strip the membrane by incubation for 20 min, 40°C, 0.4 M NaOH and repeat the hybridization using radioactively labeled T probe.
13. Compare the distances of migration of the products of partial restriction enzyme digest of the YAC with the migration of the molecular weight marker determine the positions of the restriction sites in the YAC with respect to both telomeres.

14. Compare the experimentally determined restriction map of the YAC
    with the sequence analysis-generated restriction map of the chromo-
    somal region studied.

Table 1 shows the result of this comparison obtained with the
probe C for YAC 745D12 and the enzymes *Mlu*I *Sfi*I, and *Bst*UI.
The difference between the positions of the recognition sites in the
reference sequence and the positions of the restriction sites mapped
in the YAC gives the position of the first nucleotide of the reference
sequence in the YAC. The first nucleotide of the reference sequence
is located 65 kb from the telomere C. The *Mlu*I (*see* **Fig. 3**) and *Sfi*I
(not shown) restriction sites were found at their expected positions
in the YAC. The assignment of the CpG islands has been confirmed
by indirect end-labeling of the *Bst*UI restriction sites (*see* **Fig. 3**).
Note that the resolution of the pulse-field gel electrophoresis is sufficient to distinguish the two islands accompanying the gene *BAT I*
showing 10-kb spacing, whereas the two islands accompanying the
gene *MIC A*, which show 3-kb spacing, fuse (*see* **Note 6**).

3.2. Mapping Meiotic DSBs

The DSBs are mapped in the diploid strain containing the *rad50S*
mutation. The transfer of the YAC can be performed by lithium
acetate transformation or using the *kar1* intermediate strain as
described *(2,3)*. Here YAC 745D12 has been transferred from the
wild-type strain (AB1380) to the *rad50S* strain (2850) using *kar1*
intermediate strain (2474). For detailed transfer protocol, *see* volume 54 of this series *(3)*. The haploid *rad50S* strain (2850) containing the YAC was mated with the strain YD60 of opposite mating
type carrying the *rad50S* mutation and grown on SD medium containing lysine. The resulting diploid yeast strain 2850[YAC745D12]/
YD69 was used for mapping the meiotic DSBs *(4,27,28)*.

1. Inoculate 50 mL of YPD medium with the diploid *rad50S* strain containing the YAC until 1.2–1.6 OD at 600 nm.
2. Spin down the yeast and transfer to presporulation medium YPA
   diluting the yeast 100 times and grow until stationary phase.
3. Spin down, wash with sterile H_2O, and transfer to medium SPM
   supplemented with appropriate nutrients.

4. After 6 h, spin the cells down and prepare plugs as described in Sub-heading 3.1.1. The plugs containing control DNA are prepared from yeast grown in YPD and YPA medium.

5. Run the pulse-field gel electrophoresis as above, transfer the DNA to a membrane and hybridize the DNA immobilized on the nylon membrane with the labeled probe C or T.

An example of the mapping of the sites of meiotic breakage is shown in Fig. 4, lane 1. It is apparent that the most prominent DNA breaks observed in the YAC carrying the DNA extracted from the studied chromosomal region are located respectively at 173 and 200 kb from the telomere C. These meiotic DSBs are coincident with two minisatellite-like sequences each approx 2.5 kb long, present between the MYC A and MYC B genes. We consider it likely that these sites of meiotic DSBs in yeast are also involved in the recombination events in the human germline cells, as suggested by the genetic markers' polymorphism (26) in this chromosomal region in various human populations.

4. Notes

1. Spurious breaks caused by the mitochondrial nuclease were observed in yeast cells stored in ethanol. To avoid this problem, the yeast strain containing *nuc1* mutation could be used for the mapping *(15)*. The mapping may not be reliable in the regions of YAC close to the telomere *(29)*. To map these sites, an overlapping series of YACs could be used.

2. The sequences derived from pBR322 vector have been used for constructing the YACs and remain inserted in the vicinity of the YAC telomeres. The probes used in the described experiments are C (1.543 bp *Pvu*II/*Pst*I segment of pBR322 plasmid) and T (1.691 bp *Bam*HI/*Pvu*II segment of pBR322 plasmid).

3. Formation of the spheroplasts is followed by taking up periodically 10 µL suspension of the cells, suspending in 1 mL distilled water, and measuring OD. The final OD at 600 nm of the suspension of the spheroplasts in water should be approx 10% of the initial OD. Alternatively, Glucanex® 5 U Lyticase (Fluka) can be used.

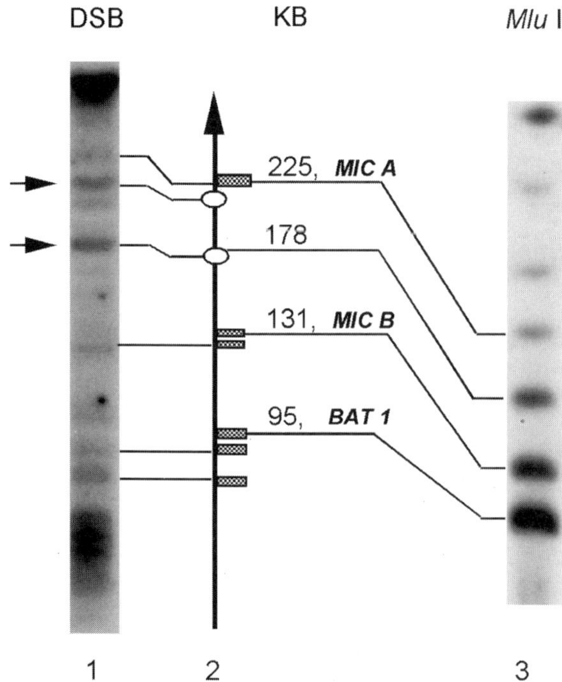

Fig. 4. Mapping of meiotic DSBs in the YAC 745D12 carrying the highly polymorphic leukocyte antigen 1 genes. The diploid yeast strain *rad50S* carrying the YAC was incubated for 6 h in the sporulation medium. The yeast DNA was subsequently isolated, size fractionated by electrophoresis, transferred to a Hybond membrane, and hybridized to a radioactive probe recognizing the telomere C. Lane 1, DNA extracted from yeast incubated in the sporulation medium. The strong DSBs are indicated by horizontal arrows; lane 2, map of the region; lane 3, DNA partially digested with *Mlu*I (ACGCGT) restriction enzyme. The positions of the bands correspond, up to the site located at 225 kb, to the positions of the restriction sites in the YAC determined using the sequence described *(25)*. The three *Mlu*I recognition sites proximal to the telomere C are located within the CpG islands (dark rectangles) accompanying genes *MIC A*, *MIC B*, and *BAT 1*. The fourth site flanks one of the microsatellites (ovals).

4. In this hypotonic buffer, the spheroplasts break but the chromatine structures are partially preserved protecting the DNA from digestion by endogenous nucleases.

5. Most of the restriction enzymes give satisfactory results when used for the digestion of the DNA embedded in the agarose blocks, although the higher concentrations and longer digestion time have to be used. Among the enzymes cutting within the CGCG tetranucleotide the *Bsh*1236 I and *Bst*UI were used without problems in the experiments described here. *See also* New England BioLabs Catalog and Technical Reference for detailed information about the digestion of DNA embedded in agarose.

6. The pulse-field gel electrophoresis in its version CHEF has a variable resolution depending on the programmable ramping of the electric field. The migration is not a linear function of the logarithm of molecular weight of the fractionated segments of the DNA. It is critical to use one or more molecular weight standards covering the whole range of molecular weights to correctly determine the molecular weight of the fractionated species of DNA.

Acknowledgments

We express our thanks to G. Symchen and D. Zenwirth for sharing with us their yeast strains and for many helpful suggestions and to J. Haber and A. Walther for critical reading of the manuscript. J.K. was financially supported by the SOCRATES students' exchange program between University Paris 7 and N. Copernicus University in Torun. The project was supported by KBN grant 6P04A00414 to G.I. by the French-Polish cooperation program POLONIUM 2001 and by grant from G.R.E.G. 520634 to J.F.

References

1. Klein, S., Zenvirth, D., Sherman, A., Ried, K., Rappold, G., and Simchen, G. (1996) Double-strand breaks on YACs during yeast meiosis may reflect meiotic recombination in the human genome. *Nat. Genet.* **13,** 481–484.

2. Hugerat, Y., Spencer, F., Zenvirth, D., and Simchen, G. (1994) A versatile method for efficient YAC transfer between any two strains. *Genomics* **22,** 108–117.

3. Spencer, F. and Simchen, G. (1996) Transfer of YAC clones to new yeast hosts. *Methods Mol. Biol.* **54,** 239–252.

4. Ira, G. (1999) Meiotic double strand breaks and nuclease hypersensitive sites in YACs carrying human DNA Ph.D. Thesis. M, Kopernik University, Torun, Poland.

5. Ira, G., Svetlova, E., and Filipski, J. (1998) Meiotic double-strand breaks in yeast artificial chromosomes containing human DNA. *Nucleic Acids Res.* **26,** 2415–2419.

6. Henthorn, P. S., Mager, D. L., Huisman, T. H., and Smithies, O. (1986) A gene deletion ending within a complex array of repeated sequences 3' to the human beta-globin gene cluster. *Proc. Natl. Acad. Sci. USA* **83,** 5194–5198.

7. Antonarakis, S. E., Boehm, C. D., Giardina, P. J., and Kazazian, H. H. Jr. (1982) Nonrandom association of polymorphic restriction sites in the beta-globin gene cluster. *Proc. Natl. Acad. Sci. USA* **79,** 137–141.

8. Treco, D., Thomes, B., and Arnheim, N. (1985) Recombination hot spot in the human β-globin gene cluster: meiotic recombination of human DNA fragments in *Saccharomyces cerevisiae. Mol. Cell. Biol.* **5,** 2029–2038.

9. Dresser, M. E. (2000) Meiotic chromosome behavior in *Saccharomyces cerevisiae* and (mostly) mammals. *Mutat. Res.* **451,** 107–127.

10. Pötter, T., Wedemeyer, N., van Dülmen, A., Köhnlein, W., and Göhde, W. (2001) Identification of a deletion on distant mouse chromosome 4 by YAC fingerprinting. *Mutat. Res.* **476,** 29–42.

11. Freudenreich, C. H., Kantrow, S. M., and Zakian, V. A. (1998) Expansion and length-dependent fragility of CTG repeats in yeast. *Science* **279,** 853–856.

12. Gendrel, C. G., Boulet, A., and Dutreix, M. (2000) (CA/GT)n microsatellites affect homologous recombination during yeast meiosis. *Genes Dev.* **14,** 1261–1268.

13. Nag, D. K. and Kurst, A. (1997) A 140-bp-long palindromic sequence induces double-strand breaks during meiosis in yeast *Saccharomyces cerevisiae. Genetics* **146,** 835–847.

14. Appelgres, H., Cederberg, H., and Rannug, U. (1997) Mutations at the human minisatellite MS32 integrated in yeast occur with high frequency in meiosis and involve complex recombination events. *Mol. Gen. Genet.* **256,** 7–17.

15. Debrauwère, H., Buard, J., Tessier, J., Aubert, D., Vergnaud, G., and Nicolas, A. (1999) Meiotic instability of human minisatellite *CEB1* in yeast requires DNA double-strand breaks. *Nat. Genet.* **23,** 367–371.

16. Mahadevaiah, S. K., Turner, J. M., Baudat, F., Rogakou, E. P., de Boer, P., Blanco-Rodriguez, J., et al. (2001) Recombinational DNA double-strand breaks in mice precede synapsis. *Nat. Genet.* **27,** 271–276.

17. Petes, T. D. (2001) Meiotic recombination hot spots and cold spots. *Nat. Rev. Genet.* **2,** 360–369.

18. Keeney, S. (2001) Mechanism and control of meiotic recombination initiation. *Curr. Top. Dev. Biol.* **52,** 1–53.

19. Lichten, M. and Goldman, A. S. H. (1995) Meiotic recombination hotspots. *Genetics* **29,** 423–444.

20. de Massy B., Rocco, V., and Nicolas, A. (1995) The nucleotide mapping of DNA double-strand breaks at the CYS3 initiation site of meiotic recombination in Saccharomyces cerevisiae. *EMBO J.* **14,** 4589–4598.

21. Cao, L., Alani, E., and Kleckner, N. (1990) A pathway for generation and processing of double-strand breaks during meiotic recombination in S. cerevisiae. *Cell* **61,** 1089–1101.

22. Kouprina, N., Eldarov, M., Moyzis, R., Resnick, M., and Lariomov, V (1994) A model system to assess the integrity of mammalian YACs during transformation and propagation in yeast. *Genomics* **21,** 7–17.

23. Avril, N., Deschavanne, P., Bellis, M., and Filipski, J. (1995) Loop-size spacing between CGCG Clusters in long segments of human DNA. *Biochem. Biophys. Res. Commun.* **213,** 147–153.

24. Guillaudeux, T., Janer, M., Wong, G. K.-H., Spies, T., and Geraghty, D.E. (1998) The complete genomic sequence of 424,015 bp at the centromeric end of the HLA class I region: gene content and polymorphism. *Proc. Natl. Acad. Sci. USA* **95,** 9494–9499.

25. Shiina, T., Tamiya, G., Oka, A., Yamagata, T., Yamagata, N., Kikkawa, E., et al. (1998) Nucleotide sequencing analysis of the 146-kilobase segment around the IkBL and MIC A genes at the centromeric end of the HLA class I region. *Genomics* **47,** 372–382.

26. Ota, M., Katsuyama, Y., Mizuki, N., Ando, H., Furihata, K., Ono, S., et al. (1997) Trinucleotide repeat polymorphism within exon 5 of the MICA gene (MHC class I chain-related gene A): Allele frequency data in the nine population groups (Japanese, Northern Han, Hui, Uygur, Kazakhustan, Iranian, Saudi Arabian, Greek, Italian) *Tissue Antigens* **49,** 448–454.

27. Baudat, F. and Nicolas, A. (1997) Clustering of meiotic double-strand breaks on yeast chromosome III. *Proc. Natl. Acad. Sci. USA* **94,** 5213–5218.

28. Ross L. O., Zenvirth, D., Jardim, A. N., and Dawson, D. (2000) Double strand breaks on artificial chromosomes in yeast. *Chromosoma* **109,** 226–234.
29. Borde V, Goldman, A. S., and Lichten, M. (2000) Direct coupling between meiotic DNA replication and recombination initiation. *Science* **290,** 806–809.

4

Genomic Imprinting and Its Effects on Genes and Chromosomes in Mammals

Yuji Goto and Robert Feil

1. Introduction

Genomic imprinting is a regulatory mechanism by which, at certain gene loci, one of the two alleles becomes repressed according to its parental origin. This chapter focuses on mammalian imprinting, which regulates not only the parental allele-specific gene expression on autosomal chromosomes: in the extra-embryonic tissues of female mouse embryos, it also determines the choice of the X chromosome to be inactivated. In this chapter, we introduce autosomal and the X chromosome imprinting in mammals and summarize the current understanding of its underlying mechanisms. Presented are examples of deregulation of imprinted genes in animal models and in human disease that arise as a consequence of deletion or acquisition of parental chromosomes. The chapter discusses the possible impact of imprinting on transgenes and artificial chromosomes and describes that in vitro manipulation and culture of cells and embryos may disrupt imprinting and can thereby lead to aberrant phenotype. The latter could be relevant relative to the application of transgenic and chromosome methodologies. Most of

From: *Methods in Molecular Biology, Vol. 240:*
Mammalian Artificial Chromosomes: Methods and Protocols
Edited by: V. Sgaramella and S. Eridani © Humana Press Inc., Totowa, NJ

the presented examples in the chapter relate to experimental work on mouse models. Because imprinting is highly conserved among placental mammals, however, emerging ideas and concepts are valid for human and ruminant model systems as well.

1.1. Genomic Imprinting at Autosomal Loci

Genomic imprinting refers to the epigenetic process by which gene activity is regulated according to parental origin. To date, some 50 autosomal genes have been found to be subject to imprinting in humans and mice. About two fifths of these are expressed from the maternally derived allele, whereas the remainder are expressed from the paternally inherited allele *(1,2)*. Most imprinted genes map to chromosomal domains that had been defined to be imprinted based on studies on mice that were uniparentally disomic or had uniparental duplication/deficiency of individual chromosomes. Through detailed developmental studies on such unbalanced contributions of maternal and paternal chromosomes, imprinted chromosomal regions have been identified *(3;* **Fig. 1***)*. One characteristic of imprinted genes is that they are organized in clusters, and there are several examples of neighboring, reciprocally expressed, imprinted genes that are coregulated. For instance, on the distal portion of mouse chromosome 7 (and on human chromosome 11q15), the insulin-like growth factor-2 (*Igf2*) gene is expressed from the paternally inherited allele only and is flanked by the tightly coregulated *H19* gene, which is expressed from the maternal allele only. *H19*, like several other imprinted genes, transcribes a nontranslatable RNA. About one in four of the identified imprinted genes produce noncoding RNAs. Some of these are in antisense orientation to other genes and might therefore be involved in transcriptional regulation *(2)*.

1.2. Imprinted X Chromosome Inactivation

X chromosome inactivation (X inactivation) is the mechanism that compensates the twofold difference in X-linked gene dosage between XX females and XY males in mammals *(4,5)*. In contrast with the allelic repression of imprinted genes, which effects genes in domains up to several mega-bases in size, X inactivation is a process that effects the majority of genes along the chromosome. Developmental studies in the mouse have established that X inactivation occurs in three waves during early embryogenesis *(5,6)*. This is depicted in **Fig. 2**. The paternally inherited X chromosome (X^P) is selectively inactivated in the first and the second wave, which occur in the trophectoderm of blastocysts at 3.5 d postcoitum (dpc) and in primitive endoderm of implanting blastocysts (4.5 dpc), respectively *(7)*. Following this imprinted type of X inactivation in the extra-embryonic lineages of the mouse, a random type of X inactivation occurs in the embryo proper during the final wave of inactivation that takes place in the epiblast at around 5.5 dpc *(8)*. X inactivation is controlled by a unique locus, the X inactivation center (*Xic*), which is essential for the initiation and spreading of inactivation along the X chromosome *(5,9)*. The *Xist* gene maps to *Xic* and is expressed from the inactive X chromosome exclusively in differentiated cells. Knockout and transgenic experiments in the mouse have shown that *Xist* expression is essential for X inactivation *(10–15)*. In female embryos, expression of the 18-kb *Xist* RNA is first detected from the paternal allele, at the four-cell stage, and this paternal chromosome-specific expression continues until the blastocyst stage *(16,17)*. Fluorescence *in situ* hybridization (FISH) has revealed that the nontranslatable Xist RNA, produced from the X^P, becomes attached to the X chromosome *in cis* in eight-cell stage embryos and in trophectoderm cells of female blastocysts *(18,19)*. Based on its expression specifically from the X chromosome that is to become inactivated, and the association of Xist RNA with the chromosome *in cis*, the nonrandom *Xist* expression from the paternal allele has been proposed to be causally involved in the preferential inactivation of the paternal X chromosome in the trophectoderm lineage and in primitive endoderm *(20)*.

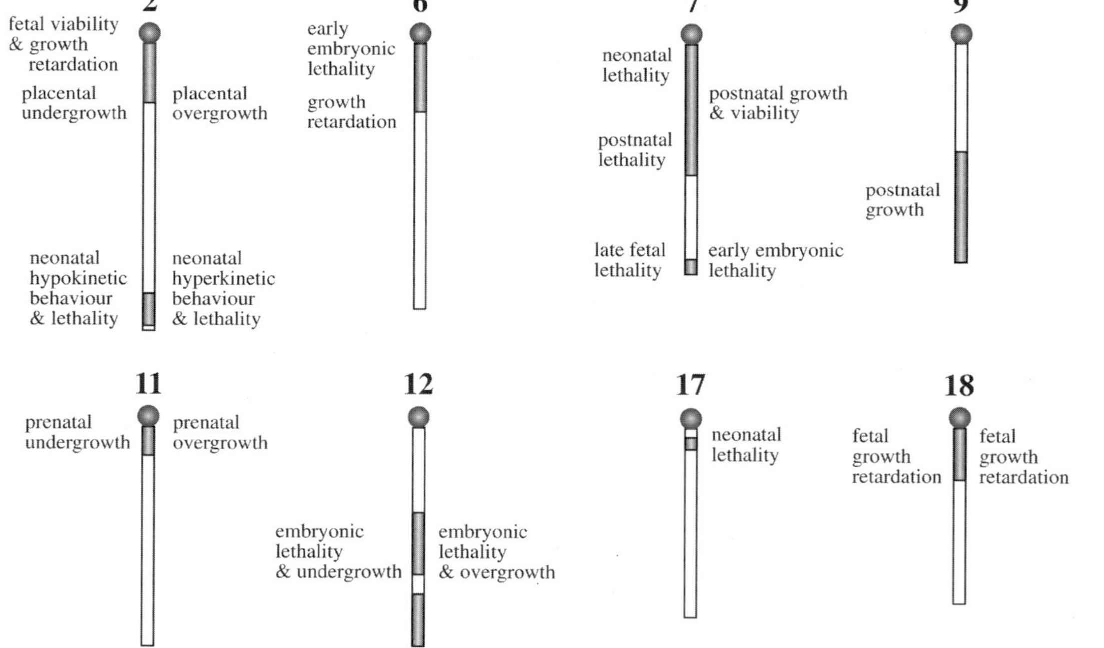

Fig. 1. Imprinted chromosomal regions in the mouse genome. Robertsonian and reciprocal translocations have been used to obtain evidence of imprinting for 11 different regions on eight autosomal chromosomes. These extensive studies *(3)* are presented on the website of the MRC Mammalian Genetics Unit (Harwell, UK: www.mgu.har.mrc.ac.uk), from which this simplified summary figure was adapted. For each imprinted region (in gray), the main phenotypic consequences of maternal disomy and paternal disomy are indicated to the left and the right of the chromosome, respectively. Additionally, the figure depicts the mouse X chromosome, which has an aberrant extra-embryonic phenotype associated with maternal disomy *(31)*.

56

2. Effect of the Parental Origin of Chromosomes

2.1. Uniparental Disomy of Autosomal Chromosomes

In mammals, both the parental sets of chromosomes are essential for normal embryonic and postnatal development. This had become apparent from experiments in the mouse in which pronuclei were transferred after fertilization of the egg, such as to produce either androgenetic (with two paternal genomes) or gynogenetic (with two maternal genomes) embryos *(21,22)*. Upon transfer into recipient female mice, such monoparental embryos were found to be severely affected in their developmental potential and died at early stages of development. This demonstrated that both a paternal and a maternal genome are required for normal development. The same observation was made in other placental mammals as well. Parthenogenetic development (with two maternal chromosomes) in sheep, for instance, leads to embryonic lethality. Similarly as in the mouse, parthenogenetic sheep embryos are smaller than control embryos and die during early fetal development, apparently because of aberrant extra-embryonic development *(23)*.

Experimental work on mice that exhibited uniparental disomy for individual chromosomes has established that several of such disomies give rise to aberrant growth and other developmental abnormalities *(3)*. In these studies, most of the uniparental disomies were obtained by intercrossing mice heterozygous for single Robertsonian translocations (nonhomologous chromosomes joined at the centromere). A summary of the main developmental abnormalities associated with maternal or paternal disomy of individual chromosomes is presented in **Fig. 1**. Evidence for differential roles of maternal and paternal copies of chromosomal regions was first obtained for the proximal portion of mouse chromosome 11 *(24)*. Newborn animals that were paternally disomic for this region (Patdi.11) were invariably larger than controls, whereas the reciprocal Matdi.11 animals were invariably smaller than normal littermates *(24)*. Mouse chromosome 7 is another chromosome with pronounced imprinting effects at three distinct regions *(3)*. At its distal-most portion (dis7), both maternal and paternal disomy give rise to embryonic death. Detailed studies showed that the Matdi.dis7 embryos (and placentas) were severely reduced in their growth, and they died at late fetal stages *(25)*. Embryos with paternal disomy for this region, in contrast, died early during embryonic development *(26)*. Embryos that were chimeric for Patdi.dis7 (mixture of normal and Patdi.dis7 cells) survived to much later stages but showed significantly increased growth *(27)*. In humans, the region that is homologous to the distal portion of mouse chromosome 7 is found on the long arm of chromosome 11 (q15). Paternal disomy and other genetic alterations at this region can give rise to the fetal overgrowth syndrome, Beckwith–Wiedemann syndrome *(28)*. Two other imprinting syndromes in humans, Prader–Willi syndrome (PWS) and Angelman syndrome (AS), map to a region on chromosome 15, which is syntenic to the imprinted domain that is on the central portion of mouse chromosome 7 *(29)*. PWS and AS are distinct neurogenetic disorders that are frequently caused by deletions in a 3–4-mb region on 15q11-13. These common deletions are similar between the two syndromes but differ in their parental origin. PWS deletions are always on the paternal allele, whereas AS deletions are on the maternal allele *(30)*. These syndromes can be associated with uniparental disomy as well, maternal disomy in the case of PWS, and paternal disomy in the case of AS.

The studies on uniparentally disomic mice and on cases of uniparental disomy in humans established that the gross phenotypes caused by maternal disomy/paternal nullisomy are different than those caused by paternal disomy/maternal nullisomy (**Fig. 1**). This indicates that both a maternal and a paternal copy are required and that there are genes in these regions that are expressed from the maternal or the paternal chromosome copy exclusively. In total, 11 imprinted autosomal regions have been identified based on the phenotypic implications of the corresponding uniparental disomies (**Fig. 1**) and almost all the imprinted genes that have been identified map to these regions.

2.2. Parent-of-Origin Effects on the X Chromosome

Confirming earlier findings *(7)*, it has been described that mouse embryos disomic for the maternally derived X chromosome (Matdi.X), such as $X^M X^M X^P$ and $X^M X^M Y$, are characterized by severely deficient extra-embryonic tissues of trophectoderm origin and die before midgestation (**Fig. 3**) *(31)*. Because $X^M X^P Y$ mice survive without problems beyond parturition, the most probable explanation of this dramatic phenotype is that two active X chromosomes in cells of the trophectoderm lineage are responsible for the extra-embryonic ectoderm deficiency in the Matdi.X embryos. Indeed, cytogenetic studies in 6.5–7.5 dpc embryos indicate that two maternal X chromosomes remain active in cells from the extra-embryonic portion of Matdi.X conceptuses. Additional data were obtained from X inactivation studies in trophectoderm cells of Matdi.X blastocysts after the first wave of X inactivation. Replication banding and Xist RNA FISH examination strongly supported the proposition that X^M is not inactivated in $X^M X^M X^P$ and $X^M X^M Y$ blastocysts at 3.5 dpc *(32)*. One explanation of these data is that the maternal X chromosome is epigenetically modified such as to remain active in the extra-embryonic lineages *(33)*.

Whether X inactivation is subject to imprinting in humans is uncertain and studies performed so far do not provide conclusive evidence for paternal X chromosome inactivation in the extra-embryonic lineages *(34)*. However, there might be individual genes on the human X chromosome that are imprinted *(35)*. This has become apparent from studies on Turner's syndrome. This syndrome in women is caused by monosomy of all or part of the X chromosome and is characterized by short stature and frequently also by social adjustment problems. The latter symptom appears to be more severe in women with deletion of the paternal X chromosome than in women with maternal X deletion *(36)*. These findings led to the proposition that there could be paternally expressed imprinted gene(s) on the X chromosomes that are involved in

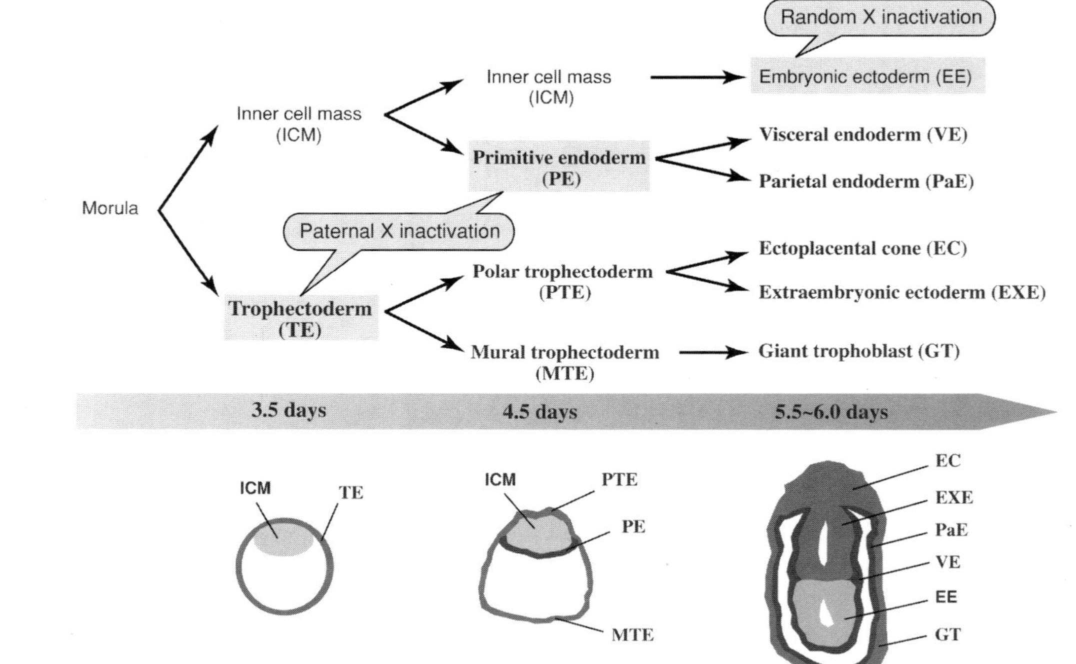

Fig. 2. Imprinted X chromosome inactivation in extra-embryonic membranes. Presented is a summary of the embryonic and extra-embryonic lineages that arise during embryonic development in the mouse. Imprinted X inactivation occurs in two subsequent waves, in the trophectoderm (at ~3.5 d post coitum) and in primitive endoderm lineage (at ~4.5 d postcoitum), respectively *(7,20)*. In all the extra-embryonic membranes that derive from these lineages, X inactivation occurs on the paternal X chromosome exclusively. Random X-inactivation, in contrast, occurs in all the embryonic tissues, that derive from the inner cell mass (ICM) cells of the blastocyst.

social and language skills. Also in the mouse, X^MO and X^PO monosomies have been compared and this revealed a difference in their early development. In particular, during pre- and early postimplantation development, X^PO embryos were found to grow faster than X^MO embryos *(37,38)*.

3. Epigenetic Mechanisms

Chromatin modifications that control the parental chromosome-specific expression of imprinted and X-linked genes should be heritable through somatic cell division and should not alter the sequence of the underlying DNA. They would also be expected to affect, directly or indirectly, the level of gene expression. The nature of epigenetic marks in imprinting has attracted a lot of interest during the last years, and methylation of DNA, chromatin modifications, and differential timing of replication have emerged as candidate mechanisms *(2,9,39)*.

3.1. DNA Methylation

Dense clusters of CpG dinucleotides are often found in the promoter or 5' regions of mammalian genes. These clusters are referred to as CpG islands and are mostly unmethylated. Exceptionally, autosomal-imprinted loci comprise CpG islands that are methylated on one of the two parental alleles. At some imprinted genes methylation is on the repressed allele, whereas at others, it is present on the expressed allele *(1,2,39)*. The latter is the case for intragenic, regulatory regions at the imprinted *Igf2* and Igf2-receptor (*Igf2r*) genes. Methylation marks at imprinted loci are essential, at least in the maintenance of imprinted gene expression. This follows from studies on mice that were deficient in DNMT1, the principal maintenance methyltransferase. Hence, in *Dnmt1* knockout mice, transcriptional reactivation of the repressed allele was observed at some imprinted loci whereas others became repressed on both the parental alleles. The activity of *Igf2*, for example, was shut off in the absence of DNA methylation, whereas the neighboring *H19*

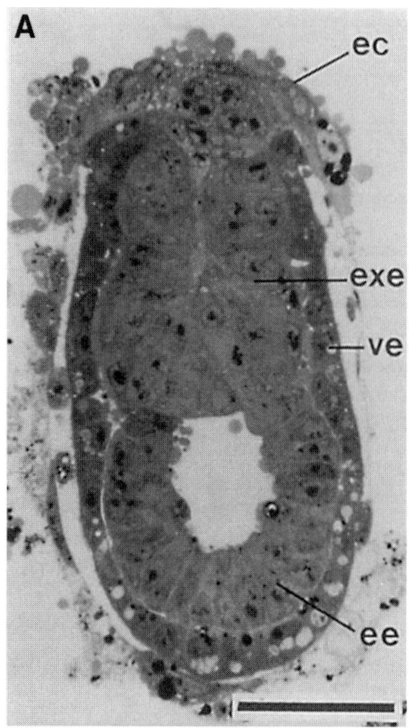

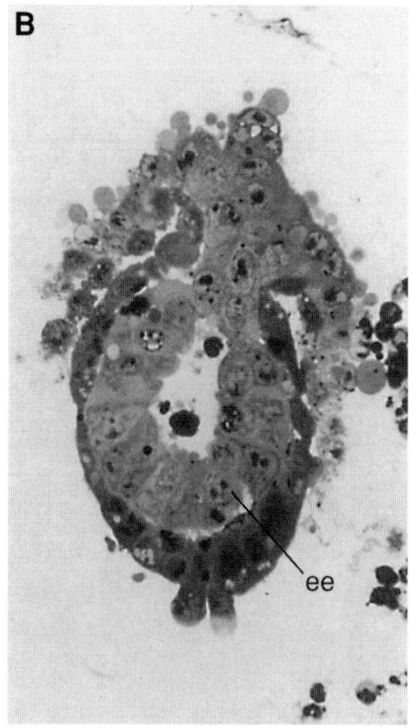

Fig. 3.

gene became expressed from both the parental alleles *(40)*. An oocyte-specific variant of DNMT1, DNMT1o, was found to be essential in the maintenance of methylation at imprinted genes as well, during a short window of preimplantation development *(41)*. At all but a few of the differentially methylated regions of imprinted loci, methylation is present on the maternally inherited allele, and for several of these maternal methylation marks it has been shown that they are established during oogenesis. The expression of a DNMT-related protein, DNMT3L, is essential for this acquisition of methylation during oogenesis *(42)*.

Relative to X inactivation, it has been found that most of the CpG islands on the inactive copy of the X chromosome are methylated, whereas the active X is unmethylated. This pattern of methylation seems to be functionally important because chemical inhibitors of cytosine methylation cause efficient reactivation of many X-linked genes. In addition, in *Dnmt1* knockout mice, the allele-specific expression of *Xist* is disrupted in the embryo proper but not in the extra-embryonic tissues *(43)*. The question remains as to whether methylation is involved in the primary establishment of X inactivation as well. Allele-specific patterns of methylation have been observed at the promoter of *Xist* in early embryos and are thought to be established during gametogenesis. It was found recently, however, that this differential methylation disappears during preimplan-

---

Fig. 3. Aberrant extra-embryonic development associated with the maternal X chromosome. Shown are histological sections of day 5.5 embryos that were stained with toluidine blue. In (**A**), a control female embryo ($X^M X^P$) with normal development of the extra-embryonic ectoplacental cone (ec), extra-embryonic ectoderm (exe), and visceral endoderm (ve). The embryonic ectoderm (ee) lineage gives rise to the embryo proper. In (**B**), a typical day 5.5 embryo with two maternal X chromosomes ($X^M X^M X^P$ or $X^M X^M Y$). Its extra-embryonic lineages are severely underdeveloped or absent; the embryo proper (ee) appears to be normal. The scale bar indicates 50 μm. This figure was adapted from Goto and Takagi, Development 1998;125, 3353–3363, with permission of the Company of Biologists Ltd.

tation development, before the actual onset of X inactivation *(44)*. It is not excluded, therefore, that methylation mark(s) at the *Xist* gene are functionally important for the initiation of the allelic transcription of *Xist*, but it seems likely that other epigenetic modifications of chromatin are involved in subsequent maintenance steps. For imprinted X inactivation, there is evidence for the involvement of modifications other than DNA methylation, because in *Dnmt1–/–* conceptuses the paternal X chromosome remains fully repressed in the extra-embryonic membranes *(43)*.

3.2. Differential Timing of Replication

The timing of replication during S-phase is thought to reflect the state of chromatin and expression of eukaryotic genes, and mammalian genes that are expressed abundantly frequently replicate earlier in S-phase than repressed genes. In the case of X inactivation, the entire inactive chromosome is highly compacted and replicates late in S-phase. Like the inactivation of genes on the X-chromosome, this late replication occurs on the paternal chromosome in the extra-embryonic tissues *(45)*. Asynchronous replication has been observed at autosomal imprinted domains as well, first by direct cytogenetic analysis at the imprinted PWS/AS region on human chromosome 15q11.2 *(46)*. Subsequent FISH studies have demonstrated differential replication between maternal and paternal alleles at the imprinted *IGF2-H19*, *IGF2R*, and PWS/AS regions in mice and humans *(47–49)*. Interestingly, asynchronous replication at some domains extends beyond the region that comprises the imprinted genes. Furthermore, imprinted domains can comprise both regions with early paternal, and regions with early maternal replication. Whether the FISH studies are entirely indicative of differential replication timing, or maybe also of structural effects on the timing of chromatid separation, is not fully resolved *(50)*. However, asynchronous replication is detected already in preimplantation embryos, and for several imprinted loci it has been found that it is reset in the male and female germ lines before the onset of meiosis. These data suggest that asynchronous replication at imprinted loci becomes established in the gametes and represents an epigenetic mark that distinguishes the parental alleles at imprinted loci *(51)*.

Which epigenetic features could underlie the asynchronous replication at imprinted loci? Possibly, the higher order organization of chromatin on the parental alleles is involved, or, the differential replication could somehow reflect specific histone modifications at the nucleosomes. It is interesting to note that allelic differences in nuclease sensitivity have been detected at X-linked and autosomal-imprinted loci, with the repressed allele being generally more resistant to nuclease digestion than the active allele *(39,52)*. Covalent modifications at the N-termini of histones alter their charge and influence local chromatin compaction and activity. Several studies now provide evidence for the occurrence of differential histone acetylation and methylation along the X chromosome, and at imprinted autosomal loci (*see* **Subheading 3.3.**).

3.3. Histone Acetylation and Methylation

At several imprinted loci, allele-specific modifications have been detected on the N-termini of the core histones H3 and H4. The differentially methylated region at the 5'-portion of the *SNRPN* gene, which is involved in PWS, has histone H3 and H4 acetylation on the expressed paternal allele only *(53)*. In **Fig. 4**, we show that the same allelic acetylation differences are found at the mouse *Snrpn* gene *(54)*. Also the promoters of the mouse *Igf2* and *H19* genes have differential H4 acetylation *(55)* and at the intragenic imprinting-control-region of the *Igf2r* gene, there are pronounced acetylation differences between the parental alleles as well *(56)*. By applying antisera specific for particular lysine residues, it has been shown that the imprinted *U2af1-rs1* gene on mouse chromosome 11 has hypo-acetylation of histone H4 on the repressed maternal allele. This maternal underacetylation was confined to the most N-terminal lysine (lysine 5), whereas at histone H3 all lysine residues analyzed were hypoacetylated *(54)*. Using a genetic approach, it was established that DNA methylation at the *U2af1-rs1* locus is linked to deacetylation of histone H3, but not H4, via a mechanism that

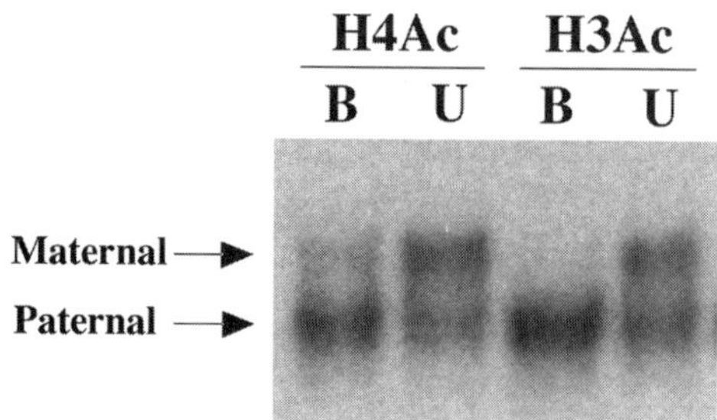

Fig. 4. Analysis of histone acetylation at the *Snrpn* imprinting control center. Chromatin was purified from interspecific hybrid mouse cells and was partially digested with micrococcal nuclease. Thus, fractionated chromatin was precipitated with antisera against hyperacetylation on histones H3 (H3Ac) and H4 (H4Ac). DNA was extracted from antibody-bound (B) and unbound (U) fractions subsequently, and a polymerase chain reaction-based approach *(54)* was used to distinguish the maternal and the paternal chromosome copies at the *Snrpn*-imprinting control region. H3 and H4 hyperacetylation is present on the paternal but not on the maternal allele *(52,54)*.

might involve association of methyl-CpG-binding domain proteins to the methylated allele *(54)*. These findings show that at imprinted loci, acetylation can be regulated differently on histones H3 and H4 and may depend on the presence of proteins that recognize methylated DNA.

Late in the X inactivation process, apparently following the transcriptional downregulation of genes, a pronounced loss of histone H4 acetylation occurs along the inactive X chromosome upon differentiation of embryonic stem cells *(57)*. Deacetylation of histone H3, at lysine 9, in contrast, is an early step in the inactivation process *(58)* and appears to be linked temporally to the acquisition of methylation at this lysine residue (*see* below). From various systems, mostly nonmammalian, it is known that other covalent modifications may occur on the N-termini of core histones as well. Methylation of lysine-9 of H3 in particular, might be relevant to the heritable repression in imprinting and X inactivation. This modification has been found to be a marker of constitutive and facultative heterochromatin. H3-K9 methylation can constitute a binding site for heterochromatin protein1, a nonhistone component of hetero-chromatin. This provides a putative mechanism whereby lysine-specific histone modifications may perpetuate states of chromatin compaction and gene repression. So far, a single study on an auto-somal imprinted locus, reported that on the human *SNRPN* gene there is H3–K9 methylation on the repressed maternal allele at its 5' imprinting-control region *(59)*. At the inactive X chromosome in mammals, methylation of H3 at lysine 9 has been detected as well. Studies on differentiating embryonic stem (ES) cells show that the acquisition of this epigenetic mark is an early event in X inactivation, which occurs after the XIST RNA paints the X-chromosome *(58,60)*.

It remains to be determined whether H3–K9 methylation occurs at chromatin associated with specific DNA sequences or whether it is equally distributed along the inactive X chromosome. This epige-netic modification has been found, however, to affect the promoters of inactive genes in differentiated cells. In addition, the H3–K9 methylation on the inactive X appears to remain present throughout mitosis and may therefore be involved in the somatic maintenance of the repressive chromatin on the inactive X *(61)*.

4. Effects on Transgenes and Artificial Chromosomes

To determine the minimal sequences required for the allele-spe-cific expression of autosomal-imprinted genes, several groups have introduced transgenes into mice. For some loci, this has enabled identification of key regulatory sequences. Most imprinted genes are organized in large clusters, however, and may share common regula-tory elements. It has been difficult, therefore, to mimic the precise expression patterns of imprinted genes from transgenic constructs. Constructs based on plasmids, cosmids, yeast artificial chromo-somes, or bacterial artificial chromosomes did frequently not give faithful imprinted gene expression in transgenic animals, even when inserted into the genome as single copies. As opposed to many nonimprinted loci, sequences of up to hundreds of kilobases appear to be required *in cis* for the correct expression of imprinted genes. Analysis of *H19* and *Igf2* expression from a 130-kb yeast artificial chromosome transgene, for example, revealed that enhancers required for their expression in mesodermal components of tissues were absent from the construct and located at greater distances *(62)*. A bacterial artificial chromosome construct that spanned no less than 315 kb and included *p57Kip2*, an imprinted gene in the same cluster as *Igf2* and *H19*, established that imprinted *p57Kip2* expression required elements that are located outside of the transgene *(63)*. A transgenic study on the allelic regulation of the *Igf2r* gene by an overlapping antisense transcript showed that critical *cis* elements are located outside of a large region that was included in the construct *(4,64)*. Another putative complication is that presence of the transgene(s) may affect the epigenetic modification and expression of endogenous imprinted genes. A dramatic effect was observed in studies on *Igf2 (65)*, in which presence of *Igf2* transgene copies led to *trans*-activation of the endogenous *Igf2* gene. This *Igf2* overexpression gave rise to fetal overgrowth and specific developmental abnormalities. In a study on a transgenic construct comprising the mouse *U2af1-rs1* locus, an inverse effect was observed *(66)*. When *U2af1-rs1* transgenes were transmitted through the male germline, this led to DNA methylation and repression on the paternally inherited endogenous *U2af1-rs1* gene (which is normally unmethylated and expressed). Although causal mechanisms are unclear, these studies establish that introduction of additional imprinting-regulating elements may have aberrant effects on endogenous loci.

5. Outlook

With delivery systems and vectors becoming more efficient, mammalian artificial chromosomes will likely be used for scientific, biotechnological, or medical purposes. Under certain conditions, however, the derivation and subsequent culture of recombinant cell lines could give rise to undesirable effects. In mice and ruminant species, it has been found that in vitro culture of cells and embryos can lead to aberrant growth and organ abnormalities at fetal stages of development *(67)*. In sheep and cattle, these phenotypic abnormalities are referred to as the large offspring syndrome. One hypothesis to account for the stochastic nature and the diversity of culture-induced phenotypes, says that in vitro culture may induce epigenetic alterations at genes and that such alterations would affect the expression of imprinted genes in particular. Recent work in mouse and rat model systems, and on early sheep embryos, provides evidence in support of this molecular hypothesis. Hence, culture of embryonic stem cells, fibroblasts, and early embryos, can lead to methylation changes at imprinting control centres *(67,68)*. These altered patterns persist throughout subsequent development and give rise to deregulated imprinted gene expression with consequences for development and growth.

The question of whether imprinting could effect mammalian artificial chromosomes gains some interest now that genes can be successfully introduced into mammalian artificial chromosomes and give rise to stable expression during many cell generations. In this chapter we summarized the complex regulation of imprinted and X-linked genes and we described that gene dosage and precise locus organization are essential for their correct expression. Because of these unique features, and its apparent sensitivity to in vitro culture and manipulation, imprinting could give rise to aberrant effects when transgenes and artificial chromosome constructs are introduced into mammalian cells and tissues.

References

1. Surani, M. A. (1998) Imprinting and the initiation of gene silencing in the germ line. *Cell* **93**, 309–312.
2. Reik, W. and Walter, J. (2001) Genomic imprinting: Parental influence on the genome. *Nat. Rev. Genet.* **2**, 21–32.
3. Cattanach, B. M. and Beechey, C. V. (1997) Genomic imprinting in the mouse: Possible final analysis, in *Genomic Imprinting* (Reik, W. and Surani, A., eds.), Oxford University Press, New York, pp. 118–145.
4. Lyon, M. F. (1961) Gene action in the X-chromosome of the mouse (*Mus musculus* L.). *Nature* **190**, 372–373.

5. Avner, P. and Heard, E. (2001) X-chromosome inactivation: counting, choice, and initiation. *Nat. Rev. Genet.* **19,**249–253.

6. Takagi, N., Sugawara, O., and Sasaki, M. (1982) Regional and temporal changes in the pattern of X-chromosome replication during the early post-implantation development of the female mouse. *Chromosoma* **85,** 275–286.

7. Takagi, N. and Sasaki, M. (1975) Preferential inactivation of the paternally derived X chromosome in the extraembryonic membranes of the mouse. *Nature* **256,** 640–642.

8. Monk, M. and Harper, M. I. (1979) Sequential X chromosome inactivation coupled with cellular differentiation in early mouse embryo. *Nature* **281,** 311–313.

9. Rastan, S. (1983) Non-random X-chromosome inactivation in mouse X-autosome translocation embryos-location of the X inactivation centre. *J. Embryol. Exp. Morphol.* **78,** 1–22.

10. Penny, G. D., Kay, G. F., Sheardown, S. A., Rastan, S., and Brockdorff, N. (1996) Requirement for *Xist* in the X chromosome inactivation. *Nature* **379,** 131–137.

11. Marahrens, Y., Loring, J., and Jaenisch, R. (1998) Role of *Xist* gene in X chromosome choosing. *Cell* **92,** 657–664.

12. Lee, J. T. and Jaenisch, R. (1997) Long range *cis* effects of ectopic X-inactivation centres on a mouse autosome. *Nature* **386,** 275–279.

13. Herzing, L. B. K., Romer, J. T., Horn, J. M., and Ashworth, A. (1997) *Xist* has properties of the X-chromosome inactivation centre. *Nature* **386,** 272–275.

14. Lee, J. T., Lu, N., and Han, Y. (1999) Genetic analysis of the mouse X inactivation center defines an 80-kb multifunction domain. *Proc. Natl. Acad. Sci. USA* **96,** 3836–3841.

15. Heard, E., Mongelard, F., Arnaud, D., Chureau, C., Vour'c, C., and Avner, P. (1999) Human XIST yeast artificial chromosome transgenes show partial X inactivation center function in mouse embryonic stem cells. *Proc. Natl. Acad. Sci. USA* **96,** 6841–6846.

16. Kay, G. F., Penny, G. D., Patel, D., Ashworth, A., Brockdorff, N., and Rastan, S. (1993) Expression of *Xist* during mouse development suggests a role in the initiation of X chromosome inactivation. *Cell* **72,** 171–182.

17. Kay, G. F., Barton, S. C., Surani, M. A., and Rastan, S. (1994) Imprinting and X chromosome counting mechanisms determine *Xist* expression in early mouse development. *Cell* **77,** 639–650.

18. Sheardown, S. A., Duthie, S. M., Johnston, C. M., Newall, A. E. T., Formstone, E. J., Arkell, R. M., et al. (1997) Stabilization of *Xist* RNA mediates initiation of X chromosome inactivation. *Cell* **91,** 99–107.

19. Panning, B., Dausman, J., and Jaenisch, R. (1997) X chromosome inactivation is mediated by *Xist* RNA stabilization. *Cell* **90,** 907–916.

20. Huynh, K. D. and Lee, J. T. (2001) Imprinted X inactivation in eutherians: A model of gametic execution and zygotic relaxation. *Cur. Opin. Cell. Biol.* **13,** 690–697.

21. McGrath, J. and Solter, D. (1984) Completion of mouse embryogenesis requires both the maternal and paternal genomes. *Cell* **37,** 179.

22. Surani, M. A., Barton, S. C., and Norris, M. L. (1984) Development of reconstituted mouse eggs suggests imprinting of the genome during gametogenesis. *Nature* **308,** 548.

23. Feil R., Khosla S., Cappai, P., and Loi, P. (1998) Genomic imprinting in ruminants: allele-specific gene expression in parthenogenetic sheep. *Mammalian Genome* **9,** 831–834.

24. Cattanach, B. M. and Kirk, M. (1985) Differential activity of maternally and paternally derived chromosome regions in mice. *Nature* **315,** 496–498.

25. Searle, A. G. and Beechey, C. V. (1990) Genome imprinting phenomena on mouse chromosome 7. *Genet. Res.* **56,** 237.

26. McLaughlin, J., Szabo, P., Haegel, H., and Mann, J. R. (1996) Mouse embryos with paternal duplication of an imprinted chromosome 7 region die at midgestation and lack placental spongiotrophoblast. *Development* **122,** 265–270.

27. Ferguson-Smith, A. C., Cattanach, B. M., Barton, S. C., Beechey, C. V., and Surani, M. A. (1991) Embryological and molecular investigations of parental imprinting on mouse chromosome 7. *Nature* **352,** 609–610.

28. Reik, W. and Maher, E. R. (1997) Imprinting in clusters: lessons from Beckwith-Wiedemann syndrome. *Trends Genet.* **13,** 293–295.

29. Cattanach B. M., Barr J. A., Evans E. P., Burtehshaw M., Beechey C. V., Leff S. E., et al. (1992) A candidate mouse model for Prader–Willi syndrome which shows an absence of *Snrpn* expression. *Nat. Genet.* **2,** 270–275.

30. Knoll, J. H.M., Nicholls, R. D., Magenis, R. E., Graham, J. M., Lalande, M., and Latt, S. (1989) Angelman and Prader–Willi syndrome share a common chromosome 15 deletion but differ in parental origin of the deletion. *Am. J. Med. Genet.* **32,** 285.

31. Goto, Y. and Takagi, N. (1998) Tetraploid embryos rescue embryonic lethality caused by an additional maternally inherited X chromosome in the mouse. *Development* **125,** 3353–3363.

32. Goto, Y. and Takagi, N. (2000) Maternally inherited X chromosome is not inactivated in mouse blastocysts due to paternal imprinting. *Chrom. Res.* **8,** 101–109.

33. Lyon, M. F. and Rastan, S. (1984) Parental source of chromosome imprinting and its relevance to X-chromosome inactivation. *Differentiation* **26,** 63–67.

34. Daniels, R., Zucotti, M., Kinis, T., Serhal, P., and Monk, M. (1997) *XIST* expression in human oocytes and preimplantation embryos. *Am. J. Human Genet.* **61,** 33–39.

35. Iwasa, Y. and Pomiankowski, A. (1999) Sex specific X chromosome expression caused by genomic imprinting. *J. Theor. Biol.* **197,** 487–495.

36. Skuse, D. H., James, R. S., Bishop, D. V., Coppin, B., Dalton, P., Aamodt-Leeper, G., et al. (1997) Evidence from Turner's syndrome of an imprinted X-linked locus affecting cognitive function. *Nature* **387,** 705–708.

37. Thornhill, A. R. and Burgoyne, P. S. (1993) A paternally imprinted X chromosome retards the development of the early mouse embryo. *Development* **118,** 171–174.

38. Jamieson, R. V., Tan, S. S., and Tam, P. P. (1998) Retarded postimplantation development of XO mouse embryos: impact of parental origin of the monosomic X chromosome. *Dev. Biol.* **201,** 13–25.

39. Feil, R., and Kelsey, G. (1997) Genomic imprinting: a chromatin connection. *Am. J. Hum. Genet.* **61,** 1213–1219.

40. Tucker K. L., Beard C., Bausmann J., Jackson-Grusby L., Laird, P. W., Lei, H., et al. (1996) Germ-line passage is required for establishment of methylation and expression patterns of imprinted but not nonimprinted genes. *Genes Dev.* **10,** 1008–1020.

41. Howell, C. Y., Bestor, T. H., Ding, F., Latham, K. E., Mertineit, C., Trasler, J. M.,et al. (2001) Genomic imprinting disrupted by a maternal effect mutation in the *Dnmt1* gene. *Cell* **104,** 829–838.

42. Bourc'his, D., Xu, G. L., Lin, C. S., Bollman, B., and Bestor, T. H. (2001) Dnmt3L and the establishment of maternal genomic imprints. *Science* **294,** 2536–2539.

43. Sado, T., Fenner, M. H., Tan, S. S., Tam, P., Shioda, T., and Li, E. (2000) X inactivation in the mouse embryo deficient for Dnmt1: Distinct effect of hypomethylation on imprinted and random X inactivation. *Dev. Biol.* **225,** 294–303.

44. McDonald, L. E., Paterson, C. A., and Kay, G. F. (1998) Bisulfite genomic sequencing-derived methylation profile of the *Xist* gene throughout early mouse development. *Genomics* **54,** 379–386.

45. Riggs, A. D. and Pfeifer, G. P. (1992) X-chromosome inactivation and cell memory. *Trends Genet.* **8,** 169–174.

46. Izumikawa, Y., Naritomi, K., and Hirayama, K. (1992) Replication asynchrony between homologs 15q11.2: Cytogenetic evidence for genomic imprinting. *Hum Genet.* **87,** 1–5.

47. Kitsberg, D., Selig, S., Brandeis, M., Simon, I., Keshet, I., Driscoll, D. J., et al. (1993) Allele-specific replication timing of imprinted gene regions. *Nature* **364,** 459–463.

48. Knoll, J. H., Cheng, S. D., and Lalande, M. (1994) Allele specificity of DNA replication timing in the Angelman/Prader-Willi syndrome imprinted chromosomal region. *Nat. Genet.* **6,** 41–46.

49. Gunaratne, P. H., Nakao, M., Ledbetter, D. H., Sutcliffe, J. S., and Chinault, A. C. (1995) Tissue-specific and allele-specific replication timing control in the imprinted human Prader–Willi syndrome region. *Genes Dev.* **9,** 808–820.

50. Bickmore, W. A. and Carothers, A. D. (1995) Factors affecting the timing and imprinting of replication on a mammalian chromosome. *J. Cell Sci.* **108,** 2801–2809.

51. Simon, I., Tenzen, T., Reubinoff, B. E., Hillman, D., McCarrey, J. R., and Cedar, H. (1999) Asynchronous replication of imprinted genes is established in the gametes and maintained during development. *Nature* **401,** 929–932.

52. Gregory, R. I., O'Neill, L. P., Randall, T. E., Fournier, C., Khosla, S., Turner, B. M., et al. (2002) Inhibition of histone deacetylases alters allelic chromatin conformation at the imprinted *U2af1-rs1* locus in mouse embryonic stem cells. *J. Biol. Chem.* **277,** 11728–11734.

53. Saitoh, S. and Wada, T. (2000) Parent-of-origin specific histone acetylation and reactivation of a key imprinted gene locus in Prader–Willi syndrome. *Am. J. Hum. Genet.* **66,** 1958–1962.

54. Gregory, R. I., Randall, T. E., Johnson, C. A., Khosla, S., Hatada, I., O'Neill, L. P., et al. (2001) DNA methylation is linked to

deacetylation of histone H3, but not H4, on the imprinted genes Snrpn and U2af1-rs1. *Mol. Cell Biol.* **21,** 5426–5436.

55. Grandjean, V., O'Neill, L., Sado, T., Turner, B., and Ferguson-Smith, A. (2001) Relationship between DNA methylation, histone H4 acetylation and gene expression in the mouse imprinted *Igf2-H19* domain. *FEBS Lett.* **488,** 165–169.

56. Hu, J. F., Pham, J., Dey, I., Li, T., Vu, T. H., and Hoffman, A. R. (2000) Allele-specific histone acetylation accompanies genomic imprinting of the insulin-like growth factor II receptor gene. *Endocrinology* **141,** 4428–4435.

57. Keohane, A. M., O'Neill, L. P., Belyaev, N. D., Lavender, J. S., and Turner, B. M. (1996) X-Inactivation and histone H4 acetylation in embryonic stem cells. *Dev. Biol.* **180,** 618–630.

58. Heard, E., Rougeulle, C., Arnaud, D., Avner, P., Allis, C. D., and Spector, D. L. (2001) Methylation of histone H3 at Lys-9 is an early mark on the X chromosome during X inactivation. *Cell* **107,** 727–738.

59. Xin, Z., Allis, C. D., and Wagstaff, J. (2001) Parent-specific complementary patterns of histone H3 lysine 9 and H3 lysine 4 methylation at the Prader–Willi syndrome imprinting center. *Am. J. Hum. Genet.* **69,** 1389–1394.

60. Peters, A. H., Mermoud, J. E., O'Carroll, D., Pagani, M., Schweizer, D., Brockdorff, N., and Jenuwein T. (2002) Histone H3 lysine 9 methylation is an epigenetic imprint of facultative heterochromatin. *Nat. Genet.* **30,** 77–80.

61. Boggs, B. A., Cheung, P., Heard, E., Spector, L., Chinault, A. C., and Allis, C. D. (2002) Differentially methylated forms of histone H3 show unique association patterns with inactive human X chromosomes. *Nat. Genet.* **30,** 73–76.

62. Ainscough, J. F., John, R. M., Barton, S. C., and Surani, M. A. (2000) A skeletal muscle-specific mouse *Igf2* repressor lies 40 kb downstream of the gene. *Development* **127,** 3923–3930.

63. John R. M., Aparicio S. A., Ainscough J. F., Arney K. L., Khosla S., Hawker K., et al. (2001) Imprinted expression of neuronatin from modified BAC transgenes reveals regulation by distinct and distant enhancers. *Dev. Biol.* **236,** 387–3899.

64. Sleutels, F. and Barlow, D. P. (2001) Investigation of elements sufficient to imprint the mouse Air promoter. *Mol. Cell. Biol.* **21,** 5008–5017.
65. Sun, F. L., Dean, W., Kelsey, G., Allen, N. D., and Reik, W. (1997) Transactivation of Igf2 in a mouse model of Beckwith-Wiedemann syndrome. *Nature* **389,** 809–815.
66. Hatada, I., Nabetani, A., Arai, Y., Ohishi, S., Suzuki, M., Miyabara, S., et al. (1997) Aberrant methylation of an imprinted gene U2af1-rs1(SP2) caused by its own transgene. *J. Biol. Chem.* **272,** 9120–9122.
67. Khosla, S., Dean, W., Reik, W., and Feil, R. (2001) Culture of preimplantation embryos and its long-term effects on gene expression and phenotype. *Hum. Reprod. Update* **7,** 419–427.
68. Dean W., Bowden L., Aitchison A., Klose J., Moore T., Meneses J. J., Reik W., and Feil R. (1998) Altered imprinted gene methylation and expression in completely ES cell-derived mouse fetuses: Association with aberrant phenotypes. *Development* **125,** 2273–2282.

5

Centromeres and Neocentromeres

Gérard Roizés, Christoph Grunau, Jérôme Buard,
Albertina De Sario, and Jacques Puechberty

1. Introduction

A functional centromere is formed by a chromosomal domain that
very often, but not always, is recognizable by a primary constriction
in metaphasic chromosomes. It is associated with a kinetochore
through which a link is established with the microtubules, which pull
the sister chromatids toward the poles of the two daughter cells during cell division. Centromeres therefore mediate chromosome segregation during mitosis and meiosis, but in ways that are relatively
different. For instance, binding of the two sister chromatids is normally destroyed at anaphase in mitosis whereas binding is maintained
in meiosis all along the first division until anaphase of the second is
reached. The same locus is, however, in charge of the two distinct
processes.

The centromeric function has been conserved during evolution
although morphology is widely varying through the species. For
instance, in the nematode *Caenorhabditis elegans* several kinetochores are dispersed along chromosomes to form the so-called
holocentric centromeres. In the same species, the trilamellar struc-

From: *Methods in Molecular Biology, Vol. 240:*
Mammalian Artificial Chromosomes: Methods and Protocols
Edited by: V. Sgaramella and S. Eridani © Humana Press Inc., Totowa, NJ

ture of the kinetochores disappears during meiosis and the contacts between microtubules and chromosomes are established through telomeric chromatin.

The centromere/kinetochore complex of higher eukaryotes is constituted by a number of components (DNA and proteins) that are more and more well defined. Paradoxically, if some centromere binding proteins are evolutionary conserved and if most centromeres contain masses of highly repeated sequences, their diversity remains a puzzling problem.

In this chapter, we essentially deal with the DNA nature of centromeres and neocentromeres in a number of species with special emphasis put on human chromosomes. We review different aspects, including the chromatin nature in which they are engaged, evolutionary considerations, their influence on the expression of genes in their vicinity, recombination, and replication timing during the cell cycle.

2. Defining the Boundaries of the Centromeric Sequences in Higher Eucaryotes Has Been a Long-Standing Challenge

Centromeres of lower eukaryotes require defined sequences that are reduced to a 125-bp stretch in *Saccharomyces cerevisiae (1)*. If in other budding yeast systems similar situations are found, centromeric DNA is considerably more complex in the fission yeast *Schizosaccharomyces pombe*, where on all three chromosomes the centromeric regions range from 40 to 100 kb, approx 7 kb of a centromeric K-type repeat and the central core being together sufficient to establish a functional *S. pombe* centromere *(2)*. Interestingly, the *S. pombe* centromere region has been shown to be composed of nonoverlapping protein domains: transcriptional silencing being mediated by different proteins at the inner or outer repeats and at the central core *(3)*; similarly, proteins located at different sites are essential for chromosome transmission and, finally, the inner-centromere and the flanking heterochromatin domains seem to be separated by boundary elements. A similar organization seemingly holds within *Drosophila melanogaster* centromere region as, for instance,

Cid-containing chromatin (Cid is the equivalent of CENP-A in man) is nonoverlapping with the chromatin that contains Hp1, a protein involved in centric heterochromatin function.

In *D. melanogaster*, the development of the X chromosome-derived Dp1187 minichromosome system has permitted the localization of the centromere to a 420-kb region that consists of two types of simple repetitive satellite DNA sequences interspersed at intervals with transposable elements and flanked on one side by an AT-rich sequence *(4)*. In spite of this, it has not been so far possible to characterize the region further and to determine which portion actually acts as the centromeric locus.

Genetic approaches have not been of any help until now to delimit the centromeric region, with the noticeable exception, however, of a mutant of *Arabidopsis thaliana*, where the standard diploid cell produces four adjoined haploid cells, a tetrad, as in yeast. By analyzing these four cells resulting from a unique gamete, recombination events could be charted through individual meioses. Scoring the segregation of centromere-linked markers allowed us to localize the centromeres of all five chromosomes *(5)*, thus providing the basis for further identification of the minimal sequence that provides centromere function.

In mammalian species, however, the situation is far from being clarified in that respect. As in all higher eukaryotes, at the primary constriction highly repeated sequences, satellite DNAs, are normally present, but attempts to delimit the centromere have been so far unsuccessful in most instances. Neither has it been possible to define clear subdomains in these regions.

Specific DNA probes and different DNA binding protein antibodies have been used to characterize the primary constriction that is most often superimposed with the several megabases of satellite DNA engaged in the centromeric permanent heterochromatic structure of animals and plants. But this approach has never allowed us to determine which portion of the region is actually representing the centromere *per se*.

The recent production of the complete DNA sequence of the human genome has not given any insight into its centromeric and pericentromeric regions because no systematic sequencing has been either possible or even attempted so far.

3. Organization of the Human Centromeric Regions

Because it is always detected at the primary constriction of all human chromosomes, α-satellite DNA has been long considered necessary to form a functional centromere. This suggestion has received strong support when artificial chromosomes have been constructed from yeast artificial chromosomes (YACs) or bacterial artificial chromosomes (BACs) containing contiguous arrays of alphoid sequences *(6,7)*, so that, with the noticeable exception of the so-called neocentromeres (*see* **Subheading 7.**), α-satellite DNA sequences are indeed normally present forming an active centromere. This does not imply, however, that any α-satellite DNA array is competent to fulfill the function.

Each centromeric region has its own structural characteristics, but they have features in common, and chromosome 21 is a good example for a description with some details (**Fig. 1**).

Centromeric regions are actually patchworks of various satellite DNA arrays (**Table 1**): α-satellite blocks and satellites 1 and 3 in chromosome 21 *(8,9)*, among which one, α-satellite DNA block I, is found at the primary constriction of chromosomes 13 and 21 when a specific probe is used in fluorescence *in situ* hybridization experiments *(10)*. Determination of the corresponding sequences on chromosomes 13 and 21 has shown that they share 99.7% identity *(11)*. At the primary constriction of each chromosome the α-satellite array is composed of tandemly arranged higher order repeats (HORs) made of a variable number of basic 171-bp repeats. This number is specific to each chromosome. Chromosome 21 HOR contains 11 times the 171-bp basic repeat unit (1.9 kb) *(12,13)*. HORs of various sizes have been characterized for most alphoid arrays of the whole human karyotype. They vary from a 2-mer (340 bp, chromosome 1) to a 34-mer (5.7 kb, chromosome Y), with intermediate sizes like 17-mer (2.9 kb, chromosome 3) or 16-mer (2.7 kb, chromosome 17), as determined by various restriction enzymes (*see* **refs.**

14 and *15* for reviews). Most of these variant arrays are specific to a chromosome or are shared between a few.

The α-satellite DNA block I of chromosome 21 varies largely in size from less than 100 kb to almost 6 Mb *(16,17)*, thus showing that a rather small subset of the large alphoid domain is necessary, and probably sufficient, to form an active centromere. Its high degree of sequence homogeneity is decreasing as one reaches the borders *(18)*. This is a general feature of all alphoid blocks analyzed so far in detail. For instance, on the p side of chromosome 16, the α-satellite sequences at the junction with non-α-satellite show total absence of higher order structure *(19)* and no closer phylogenetic relationship to the chromosome 16 specific HOR α-satellite than to that of any other chromosome *(20)*.

4. Nature of the α-Satellite Sequences Competent to Form a Centromere

In spite of the absence of a precise definition of the molecular nature of a human centromere, it is now accepted that α-satellite DNA sequences are indeed necessary to seed it in normal conditions, although evidence has accumulated to indicate that the sequence *per se* is not the factor that defines the centromeric locus. The question is, however, still open to determine which part, if any, of the several megabases of α-satellite DNA present at each centromere is the most competent for the function. α-satellite DNA is common to all primates and has been continuously changing during evolution. When analyzed in details, one can distinguish, depending on the number and the nature of the 171-bp basic repeats that have formed them, several so-called suprachromosomal families *(21)* that, in turn, gave rise to HORs in the human karyotype. These correspond to new arrays (recently amplified, HOR$^+$) with a homogeneity of 95–99% between the HORs of a particular chromosome. Old arrays (anciently amplified, HOR$^-$) often coexist with the former, but they are unable to maintain homogeneity between the copies and, therefore, HORs are not observable by restriction analysis. Most reports in the literature make us think that the new arrays are indeed present at

p arm
q arm
TPTE
BAGE2
ORLP1
TTLL1P
EIF3S5P
CEN
PRED65
PRED3
BAGEP
PRED4
ORLP2
CNN2P
C21orf15
CYP4F3LP
NF1L1P
PRED5
RBM11
PRED6
STCH
SAMSN-1
POLR2CP
NRIP1
0 0.1 0.2 0 0.1 0.2 0.3 0.4 0.5 0.6 0.7 0.8 0.9 1.0 1.1 1.2 1.3 1.4 1.5 1.6 1.7 1.8 1.9 2.0 Mb
L H1 L
L
H1
L
α satellite
satellite 1
satellite 3

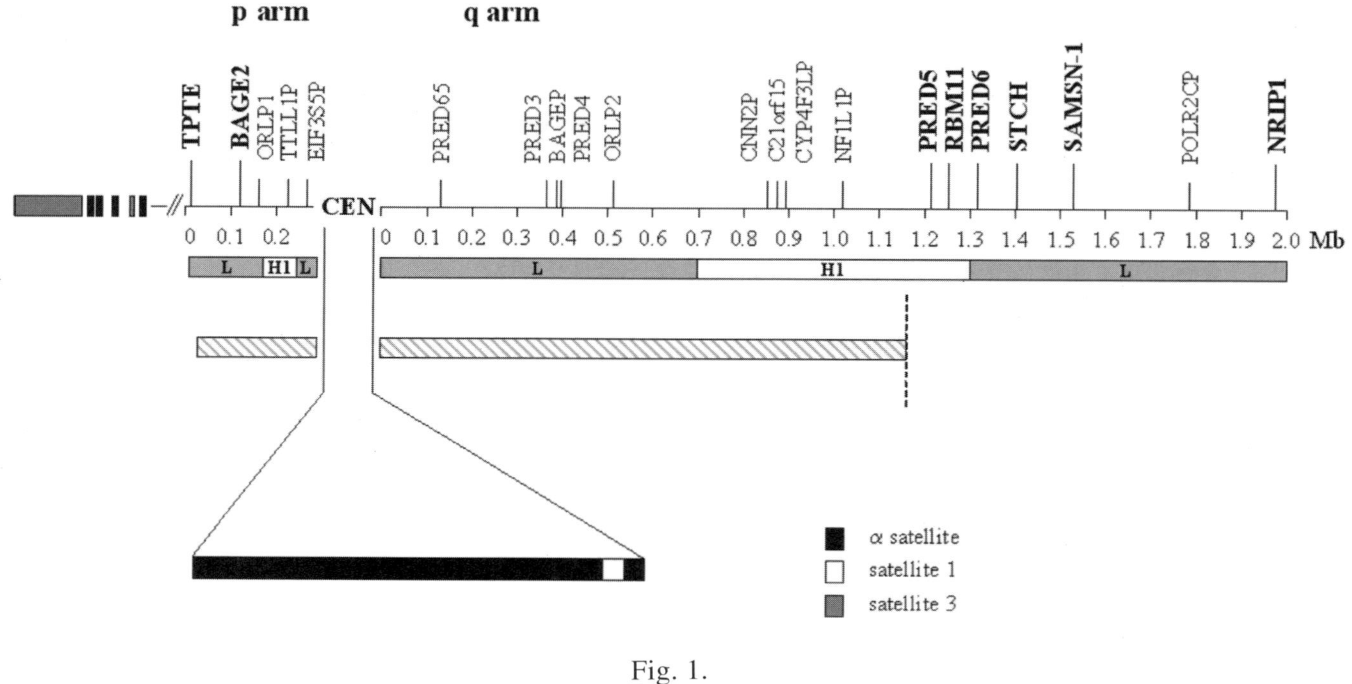

Fig. 1.

Table 1
Non-α-Satellite DNA Sequences of the Human Genome

Name	Size of the repeat unit (bp)	Detected on chromosomes
Satellite 1	42	3, 4, 13, 14, 15, 21, 22
Satellites 2 and 3	5	Probably all
β-satellite (or *Sau*3A repeat)	68	1, 9, 13, 14, 15, 21, 22, Y
γ-satellite	220	8, X
48-bp repeat	48	21, 22, Y (Possibly others)

the primary constriction of all chromosomes and therefore include the domain(s) competent to form an active centromere. This has received recent support by two independent approaches. First, Ikeno

---

Fig. 1. Chromosome 21 centromeric and juxtacentromeric regions. On both sides of 21cen are indicated the principal features characterizing the p- and q-arms few first hundreds of kilobases. Most DNA sequences belong to GC-poor L isochores *(90)* alternately with shorter GC-rich H1 isochores. At 21cen is found a long α-satellite DNA array as well as a rather short satellite DNA 1 block, the localization of which being not clearly established as it localized on the q arm close to 21cen *(8)*, whereas the published DNA sequence of human chromosome 21 *(25)* begins with 31 kb of uninterrupted α-satellite DNA sequences. We have therefore provisionally placed the satellite 1 block as if it was embedded within alphoid sequences because this fits better with both types of results. On the p side of 21cen, at a distance that has not yet been determined, is found a patchwork of α-satellite and satellite 3 DNA blocks. Again, there is no clear evidence in the literature *(8,9)* of their relative arrangements. Most of the DNA sequences belonging to the pericentromeric regions on both sides of 21cen are made of duplicated genomic sequences (here represented by long hatched rectangles) with a high degree of sequence identity shared with other chromosomes, in particular with the other acrocentrics. (For a detailed analysis, *see* Figs. 31 and 32 of **ref. *91*.**) On top are indicated the gene fragments, pseudogenes, and the genes (in bold characters) that have been characterized. The vertical dotted line marks the putative boundary between hetero- and euchromatin.

et al. *(7)* constructed a human artificial chromosome with a YAC containing about 100 kb of α-satellite sequences of the chromosome 21 block I (HOR⁺), the same attempt with another YAC containing the same quantity of α-satellite of chromosome 21 block II (HOR⁻) being unsuccessful. Even more demonstrative in that respect is the work by Schueler et al. *(22)*: they deleted the flanking DNA of the chromosome X α-satellite DNA (HOR⁺), the deletion eliminating more diverged α-satellite DNA sequences (old repeats, HOR⁻): the mitotic stability of the chromosome was maintained, showing that the highly homogeneous α-satellite DNA included the sequences able to form an active centromere.

5. Junctions Between α- and Non-α-Satellite DNA Sequences

The sequence of the human genome has not given much insight in the nature of the boundaries between alphoid and nonalphoid sequences, strongly supposed to represent a transition between heterochromatin and euchromatin. Most of the data that have been collected so far originate from the few sporadic efforts devoted to these regions. Difficulties in obtaining clones of nonrearranged alphoid DNA stretches, in assembling the resulting sequences and, moreover, the fact that paralogous sequences with high sequence identity are shared between numerous chromosomes, have indeed limited the efforts to a small number.

To illustrate these difficulties, it is interesting to note that the proximal 10q region, which is the most thoroughly analyzed and sequenced one *(23)*, does not include this junction which has however been obtained in a few chromosomes, mainly 16 *(24)*, 21 *(25)*, and X *(22)*.

At the 16p11 boundary, approx 92 kb of α-satellite DNA have been sequenced. They contain 4 L1 retrotransposed elements and belong to the D16Z2 α-satellite locus, though they failed to show any evidence of the 1.7-kb HOR structure found within the core of D16Z2 *(24)*.

On chromosome 21, 31 kb of alphoid sequences, including one L1 element, are found just before the beginning of the q arm *(25)*. Again, there is no evidence of the 1.9-kb HOR of D21Z1. Another junction, presumably on the p side, has been analyzed by Mashkova et al. *(26)*. Unfortunately, only the alphoid part of the junction has been published. Sequence comparison between the 171-bp repeats of the 18 kb of assembled alphoid sequences showed that they were highly diverged from the alphoid core and that they were interrupted by 3 L1 elements.

The p side of chromosome X has been the most thoroughly analyzed in that respect. Within the alphoid array of the primary constriction (DXZ1), two types of α-satellite sequences are detected. They both share the same 2.0-kb HOR, but type 1 repeats show 1 to 2% divergence between copies whereas type 2 repeats have substantially higher and higher divergence with increasing distance from type 1 array. Immediately distal on Xp are type 3 DXZ1 repeats which have only 85% sequence identity with type 1 DXZ1; they present no evidence of HOR and their sequence identity drops as physical distance increases further. Interestingly, L1 elements have retrotransposed within type 3 with a frequency much higher than within types 1 and 2 repeats, only one L1 being detected within the 2–3 Mb of the HOR$^+$ array.

As noticed by Laurent et al. *(27)*, L1 elements are particularly abundant within alphoid sequences, probably thanks to the presence of seven potential specific target sites of insertion within each 171-bp repeat. They are, however, almost exclusively found in abundance, as in chromosome 5 *(28)* or in the above example of chromosome X, in alphoid sequences that are quite divergent from the canonical alphoid domain of the chromosome in which they have retrotransposed. An additional characteristic of these L1s is that they often exhibit a high frequency of presence/absence polymorphism, contrary to those detected in the chromosome arms *(29)*. It has been proposed that they could shape to some extent the centromeric regions *(29)*.

6. Pericentromeric Regions of Human Chromosomes

The centromeric region of each human chromosome is embedded within a larger region of constitutive heterochromatin, the borders of which are again difficult to define precisely, the complete

DNA sequence of the human genome having not shed much light on their nature and structure. They are, however, outside the centromeric domain itself as shown on chromosome X, where they could be deleted with no effect on the centromeric activity itself *(22)*.

The striking feature of most human pericentromeric regions analyzed so far is that they often share large blocks (>1 kb) of duplicated genomic sequences with a high degree of sequence identity (>90%). This feature is shared, to a lesser extent, with the subtelomeric regions of human chromosomes *(30)*. The number of pericentromeric regions that bare such paralogous duplicated regions is quite large. It will become larger when the specialized efforts aiming at targeting the missing pericentromeric regions of the human genome are successful.

As suggested by Horvath et al. *(31)*, these duplications have arisen from a two-step process. In the first, genomic segments duplicatively transpose to an ancestral pericentromeric region, thus generating a mosaic of duplicated segments originating from various loci. Then, different subsets of the ancestral pericentromeric region spread by duplication to other pericentromeric regions.

Pericentromeric regions are moreover prone to genomic instability, with syndromes as consequences in some instances. This is exemplified by the interstitial duplications associated with congenital malformations in which half of all chromosomal regions involved occurred within regions relatively close to them *(32)*.

The question of whether these heterochromatic regions are junk or valuable has posed a problem for a long time. Several observations have already shown that heterochromatin variations could affect the phenotype as, for instance in *Drosophila*, where the deletion, or addition, of heterochromatic regions affect the adult phenotype *(33,34)*.

As shown on **Fig. 1**, most of the sequences found in close proximity to 21cen on both sides are made of interchromosomal duplications which, for a large part, consist of repetitive sequences intermingled with retrotransposed pseudogenes and gene fragments *(35)*. It is also noteworthy that at least 40 genetic loci are estimated to be present in the *D. melanogaster* heterochromatin *(36)*, where their expression is heterochromatin-dependent because, when translocated into euchromatin, they are silenced *(37,38)*. It is therefore tempting to search for real genes in close proximity to human centromeres in spite of a compact structure that generally implies gene silencing. Genes indeed have been found, in limited numbers however so far, in the juxtacentromeric region of chromosome 21: *TPTE* (or *transmembrane phosphatase with tensin homology*) *(39)* and *BAGE2* (*B melanoma antigen2*) *(39a)*, which map to the short arm of chromosome 21. Both genes have two common features. First, they belong to families whose members map to the juxtacentromeric regions of different chromosomes. Second, their expression is restricted: *TPTE* is exclusively expressed in testes, whereas *BAGE* genes are expressed in testes and different cancer cells. It is tempting to suggest that the expression profile restricted to a few tissues and/or to a specific stage of development is a general feature of the genes located within such a heterochromatic environment. So far, no biological function has been attributed to *TPTE* nor to *BAGE*.

7. Inactivated Centromeres and Neocentromeres

The situation considered above is the normal status of a centromere made of α-satellite DNA sequences, embedded in a larger domain of heterochromatin often consisting of duplicated segments of genomic DNA shared by several chromosomes.

Variant situations are, however, encountered in which a rearranged chromosome has either acquired a supplementary centromere or lost its normal one. Occasionally, the variant chromosome is maintained and transmitted normally.

When an abnormal chromosome contains two regions of centromeric DNA (isocentric or dicentric), α-satellite DNA in humans, aberrant segregation may occur with consequences like chromosome breakage and/or chromosome loss. Stably transmitted dicentric chromosomes are, however, observed in humans and in other species. This stability either results from the inactivation *(40)* of one of the two centromeres, from the coordination of their activities *(41)*, or even from their alternate inactivation *(42,43)*. The mechanisms by which stability of dicentrics is ensured are unknown. Proteins essential for the function, such as CENP-A, CENP-C, or CENP-H, are, however, not detected at the inactivated centromere whereas others, like CENP-G or CENP-B, are found at both *(44)*.

During the course of a routine diagnostic cytogenetic analysis, a stable marker chromosome totally devoid of α-satellite DNA *(45)* was analyzed with a primary constriction localized within 10q25.3. Since then, a number of such activated so-called latent centromeres have been described, to which the name neocentromere has been given.

A detailed analysis of some of them has allowed us to better understand which factors could contribute to specify a locus to become an active centromere. The most thoroughly studied locus is the one found at 10q25.3. It covers 80 kb of DNA, the sequence of which has been determined and it has revealed no difference with that of the normal 10q.25.3 locus *(46)* except for only a few single nucleotide polymorphisms. This result is in strong support of an epigenetic mechanism in the neocentromerization process. One cannot, however, formally conclude that the sequence, or rather more subtle sequence motifs, either at a normal centromere or at a neocentromere, is of no importance in this process. First, examination of a set of 40 neocentromere-containing chromosomes suggests that 13q32, 13q21, 15q, and 3q have an increased propensity for neocentromere formation whereas chromosomes 18 and 21 are avoided *(47)*. Second, a comparison between the yeast 125-bp centromeric DNA, primate α-satellite DNA, and the human chromosome 10q25.3 neocentromere sequences showed that they have some features in common: AT-rich stretches and dyad symmetries *(48)*. Finally, it is possible also that bend ability could be a common trait to all centromeric sequences *(49)*.

A novel chromatin immunoprecipitation and array analysis has permitted to better characterize the 10q.25.3 region. The 80 kb are thus shown to be part of a larger 700-kb domain identified by immunoprecipitation with CREST antisera (or calcinosis, Raynaud's phenomenon, esophageal dysmotility, sclerodactyly, telangiectasia), a 330-kb CENP-A binding domain being identified within the same 700 kb with an anti-CENP-A-specific antibody *(50)*. The DNA sequences that are involved in the neocentromeric function are therefore larger than expected from earlier studies. The same approach identified a 460-kb CENP-A binding neocentromere DNA at 20p12 *(51)*.

As normal centromeres, neocentromeres show identical distribution patterns of >20 functionally important kinetochore-associated proteins, with the exception, however, of CENP-B *(52)*, which is now known to be dispensable *(53)*. Neocentromeres have also been occasionally described in *Drosophila (54)* and plants *(55)*. It is interesting to stress here that neocentromeres can also function during meiosis, as shown by a fully functional human neocentromere transmitted through three generations *(56)*.

Differences in behavior between centromeres and neocentromeres are therefore limited. Normal centromeres are involved in a heterochromatic structure contrarily to neocentromeres. Moreover, the 10q25.3 neocentromere was capable of producing minichromosomes by telomere-associated chromosome truncation as normal centromeres do, but, by contrast, it was unable to initiate the construction of neocentromere-based human functional artificial chromosomes by two different in vitro assembly approaches *(57)*. Replication timing might also be different between normal centromeres and neocentromeres (*see* **Subheading 9.**).

8. Centromere, Recombination, and Nondisjunction

The origin of trisomy has been analyzed for several chromosomes in humans. As for *Drosophila* and yeast as model organisms, both the number of recombinational events and the location of the exchanges have important effects on the rate of chromosome nondisjunction *(58)*. Although differences in behavior are to be noticed for these chromosomes (2, 7, 15, 16, 18, 21, 22, X and Y) in which nondisjunction has been analyzed, chromosome 21 is the most thoroughly studied in that respect and shows interesting characteristics more or less representative of all. Most nondisjunctions occur during the first meiotic division (M I) of the female lineage, a large proportion being achiasmate. Chromosome 16, however, is an exception, the nondisjoining chromosomes exhibiting one or two chiasmas but are more distally located, a 20-fold reduction of recombination in the proximal regions of chromosome 16 in trisomy-generating meioses being observed *(59)*. Although less pronounced, this bias is also observed in chromosome 21 trisomic meioses in which at least one chiasma is observed. If nondisjunction occurs during M I, one finds a strong shift in location of the chiasmas toward the telomere, 80% of all single exchanges being located in the telomeric third of the chromosome. If nondisjunction occurs during M II, from which no chromosome results from an achiasmate M I, those with a single chiasma revealed a shift in location toward the proximal third of the chromosome, covering the pericentromeric region.

It appears therefore that exchanges too close or too far from the centromere increase the risk for nondisjunction. The answer to the question of how this correlates, and if so how, with the phenomenon known as centromere-associated repression of recombination has still not been understood.

Recombination has been indeed considered for a long time to be reduced in the centromeric regions of chromosomes. This finding has recently been assessed by direct estimation of genetic distances around centromeres of chromosomes X *(60)* and 5 *(28)*. This has been possible because pericentromeric polymorphic genetic markers had been characterized for the two chromosomes, which is not the case for the other chromosomes, because of the complex structure of pericentromeric regions, so that it has not been possible so far to determine whether the repression of recombination around centromeres is similar for all. Preliminary results in this laboratory however indicate that this could be different at least for chromosome 21 where in the CEPH panel a relatively high recombination rate in the vicinity of 21cen has been evidenced in normally segregating chromosomes (A. M. Laurent, G. Roizés, and J. Buard, unpublished results).

9. Replication Timing of Centromeres

Chromosomal domains replicate at specific times during S-phase, which often correlates with gene activity. As a general rule, active loci replicate early, although inactive (heterochromatin), are late replicating *(61)*. Very late replication of centromeres has long been thought to be important for centromeric function in that sister chromatids would remain attached until replication of the whole genome is completed *(62)*. Nevertheless yeast centromeres have been shown to replicate in early S-phase *(63)* as apparently also those of *S. pombe*.

Recently, several reports have dealt with the problem of centromere replication timing. Contradictory results have been published with *Drosophila*. Ahmad and Henikoff *(64)* have argued that, in contrast with expectation, centromeres replicate as isolated domains early in S-phase and remain sequestered within surrounding heterochromatin until it replicates late. Replication would thus participate in centromere maintenance. This has been contradicted by Sullivan and Karpen *(65)*, who claimed that replication timing does not determine centromere identity by showing that centromeric chromatin in *Drosophila* is replicated at different times from mid-to-late S-phase.

Similar analyses are, however, extremely difficult to perform in mammals, although the bulk of α-satellite DNA has indeed been shown to be replicated in mid-to-late S-phase *(66,67)*, the same holding true for the mouse satellite DNA *(68)*. It is also interesting to note that in barley all centromeres seem to replicate in a coordinated fashion at mid S-phase *(69)*. However, as a minority of the satellite DNA sequences in each chromosome could be actually engaged in the centromere *per se*, this does not exclude that its replication could be initiated at a different time from the bulk. Indeed, a tiny portion of α-satellite DNA is replicated as early as the very beginning of S-phase, whereas the vast majority as estimated in the light cycler is replicated late and very late, including G2 (G. Roizés and C. Laird, unpublished results). This implies that several replication origins scatter the long centromeric arrays of α-satellite DNA and that the centromeric portion could be replicated at a specific time.

Neocentromeres are offering again a better opportunity to analyze replication timing more precisely. A detailed analysis of the 10q25.3 neocentromere shows that two zones surrounding the 330-kb CENP-A binding domain exhibit significantly delayed replication times compared to their normal chromosome 10 counterparts, in thus mimicking a so-called heterochromatic behavior *(50)*.

It is therefore possible that the alphoid arrays of human centromeric regions are initiated for replication at different times during S-phase as in the above model of the 10q25.3 neocentromere, which could mean that replication timing is indeed a factor, among others, participating to the centromerization of an alphoid locus. Contrary to normal nucleosomes where replication and histone synthesis are simultaneously progressing, centromeric nucleosomes are not assembled in synchrony with centromeric replication because CENP-A is synthesized during G2 *(70)*. Consistent with these results, Henikoff et al. *(71)* have proposed a sequestration model in which the centromere is replicating earlier than the surrounding heterochromatin, which would prevent histone H3 from being deposited on centromeric DNA whereas CENP-A would be free to reach its centromeric target.

Finally, it is also of interest to note here that asynchronous replication of homologous α-satellite on four different human chromosomes (10, 11, 17, and X) could be associated with chromosome nondisjunction *(72)*.

10. Evolutionary Considerations

The repeats present at the centromeres of closely related species are extremely variable in sequence, showing that they are among the most rapidly evolving DNA elements of the higher organisms' genomes. This is particularly striking when comparative mapping of human alphoid sequences is performed between great apes using *in situ* hybridization *(73)*. In this study, the authors investigated the evolutionary relationship between the centromeric regions of the common chimpanzee (*Pan troglodytes*), the pygmy chimpanzee (*Pan paniscus*), and the gorilla (*Gorilla gorilla*). They used a panel of 27 different human alphoid probes belonging to different suprachromosomal families. Surprisingly, the vast majority of the probes did not recognize their corresponding homologous chromosomes, although alphoid sequences were indeed present at the centromeres of all chromosomes as in the human karyotype. This shows that an extremely rapid rate of change is affecting the alphoid sequences during evolution.

They then extended their analysis by looking at the primate phylogenetic chromosome IX in 9 primates, including *Pan troglodytes*, *Pan paniscus*, and *Gorilla gorilla* and, among others, *Homo sapiens*. Not only are the sequences of their centromeres quite different from one species to the other, but, in addition, centromere repositioning is independent of their flanking chromosomal markers *(74)*.

Recently, Ventura et al. *(75)* have provided hints to understand centromere repositioning during evolution. In this study, they compared the X chromosome among mammals because it has conserved a strict marker order, including centromere, between felines (*Felix cattus*) and humans. This situation was ideal for testing centromere repositioning because they could investigate the order of the same markers in two species where the X chromosome morphology is different. The chromosome X is telocentric in *Eulemur macaco* (EMA) and almost metacentric in *Lemur catta* (LCA). Despite this extreme difference in centromere position, the chromosome X of these two species shows no marker order discrepancy. Centromere repositioning, which occurred at some stage of the evolution, was not therefore caused by a pericentric inversion between humans (HSA) and LCA, nor to centromere transposition between LCA and EMA. Neocentromere emergence seems instead to be the most likely explanation of the centromere repositioning on the X of these species.

If this vision of the emergence of new centromeres during evolution is confirmed in the future, a number of points still remain to be explained. How do heterochromatic material accumulate at the locus where a neocentromere emerges? Does this result necessarily from the simultaneous and gradual accumulation of tandem repeats (satellite DNA) at this locus? Or is this accomplished by the sudden insertion of satellite DNA from an already formed "normal" centromere?

A parallel can be made between rapid satellite DNA changes and those observed in the centromere-specific H3 histone. It is indeed also changing rapidly between species contrary to its normal H3 counterpart, which is well conserved as the other histones that are the most highly conserved proteins of eukaryotes. N-terminal tails bear little resemblance to those of H3 in all species so far examined, both in length and sequence similarity. The nucleosomal core region is also more divergent. This is particularly striking when Cid of *Drosophila* is analyzed in five species of the *D. melanogaster* subgroup. It is therefore postulated that centromeric H3-like proteins follow an adaptative evolution in concert with the rapid changes observed in satellite DNA at *Drosophila* centromeres *(76)*. Obviously, if true, this holds for the other species.

To account for the rapid changes in satellite DNAs, and their fixation into populations, most authors put forward the slow accumulation of unequal exchanges between sister chromatids during mitosis. However, this is essentially based on computer simulations *(77,78)*. Others have opposite opinions on this matter *(79,80)*. An alternative mechanism could be a sudden amplification of a small subset of centromeric tandem repeats formed originally by replication slippage and to some extent by unequal exchange as suggested by some reports *(29,81)*. This mechanism takes more account of the extreme variations in size of the α-satellite DNA arrays found for instance at the centromere of chromosome 21 *(82)*. Such a large-scale amplification process would be in addition compatible with the suggestion made by Henikoff et al. *(72)* that the strength of a centromere could be selected by meiotic drive because such a sudden amplification process would provide at once a variant satellite array better adapted to the changes introduced in the centromere-specific H3 histone.

11. Conclusions and Perspectives

Despite the accumulation of a large amount of data describing the DNA sequences and the proteins present at the centromere of a large panel of higher organisms, it has not yet been possible to draw clear conclusions about the determinants that give the information to a locus to become a centromere and maintain it through cell divisions. They involve both DNA sequence relative specificity and epigenetic factors.

The DNA repeats present at the centromere of closely related species are extremely variable in sequence. Despite this, some features are common to most. Satellite DNAs are generally AT-rich (with exceptions as in Bovidae and Capridae) with a basic repeat that more or less reflects the nucleosomes' sizes in most organisms (with again exceptions as in *Drosophila*). Their repeating units exhibit also a conserved pattern of bending. Other not yet detected subtle sequence arrangements also exist which could be shared with neocentromeres.

The quantity of satellite DNA at each centromere is also extremely variable between individual homologous chromosomes, as a consequence of very efficient mechanisms of contraction, expansion or *de novo* creation of satellite DNA arrays at centromeres. This is in itself a puzzling problem that is far from being understood.

The recent discovery and characterization of several neocentromeres, or nonsatellite-containing centromeres, has allowed us to speculate that non-DNA-sequence determinants could maintain centromeres through the generations. Among the epigenetic factors, which are thus envisaged, one protein, CENP-A in humans, Cse4p in *S. cerevisiae*, HCP-3 in *C. elegans*, Cid in *D. melanogaster*, and SpCENP-A in *S. pombe*, is considered as having a central role in the centromerization process. It is exclusively centromeric where it replaces the normal H3 histone in nucleosomes. It has a significant affinity for the centromeric DNA sequence or structure *(83)*. CENP-A, however, might not be the unique determinant of centromere maintenance as other proteins are involved in the edification and stabilization of the centromere/kinetochore structure. These centromeric proteins, including non-H3 histones, are also subject to modifications (the so-called histone code hypothesis) *(84)* that could play an important role in the specification of centromeres. Histone methylation, for instance, can generate binding sites for heterochromatin-associated proteins *(85)* that, in turn, could influence centromere activity.

DNA methylation could also be an important determinant in centromere activity. Although it is mainly involved in gene inactivation, it also affects centromeric DNA, particularly satellite DNAs. Contrary to tissue-specific expressed genes, which are rather methylated during gametogenesis and the early stages of embryo-genesis, satellite DNAs are undermethylated in the same tissues and become heavily methylated in adult tissues *(86,87)*. Among the diverse methyltransferases that have been characterized so far, DNMTB3B appears to be specialized for the methylation of specific compartments of the human genome. In the ICF syndrome (i.e., immunodeficiency, centromeric instability, facial abnormalities), the corresponding gene is mutated causing demethylation of the classic satellite DNAs (satellites 2 and 3) of chromosomes 1, 9, and 16, which in lymphocytes is cytogenetically visible by elongation of juxtacentromeric heterochromatin and by the formation of metaphasic multiradiate chromosomes *(88)*. It is possible therefore that methylation of the satellite DNA present at the centromere itself could be of importance also for the centromeric activity. In that respect, it is interesting to note that in the mouse *Dnmt3b* mutants have a limited effect on the demethylation of the juxtacentromeric major satellite DNA but affects the centromeric minor satellite DNA, which is substantially demethylated in the same *Dnmt3* mutant *(89)*.

The conjunction of several factors might therefore be important for the specification of a locus to become a centromere and for maintaining it through the generations. It would be surprising, however, that each would have equal importance in all metazoan species.

References

1. Clarke, L. (1998) Centromeres: Proteins, protein complexes, and repeated domains at centromeres of simple eukaryotes. *Curr. Opin. Genet. Dev.* **8,** 212–218.
2. Baum, M., Ngan, V. K., and Clarke, L. (1994) The centromeric K-type repeat and the central core are together sufficient to establish a functional *Schizosaccharomyces pombe* centromere. *Mol. Cell Biol.* **5,** 747–761.
3. Partridge, J. F., Borgstrom, B., and Allshire, R. C. (2000) Distinct protein interaction domains and protein spreading in a complex centromere. *Genes Dev.* **14,** 783–791.

4. Sun, X., Wahlstrom, J., and Karpen, G. (1997) Molecular structure of a functional *Drosophila* centromere. *Cell* **91,** 1007–1019.

5. Copenhaver, G. P., Nickel, K., Kuromori, T., Benito, M.-I., Kaul, S., Lin, X., et al. (1999) Genetic definition and sequence analysis of Arabidopsis centromeres. *Science* **286,** 2468–2474.

6. Harrington, J. J., Van Bokkelen, G., Mays, R. W., Gustahaw, K., and Willard, H. F. (1997) Formation of *de novo* centromeres and construction of first-generation human artificial chromosomes. *Nat. Genet.* **15,** 345–355.

7. Ikeno, M., Grimes, T., Nakano, M., Saitoh, K., Hoshino, H., McGill, N. I., et al. (1998) Construction of YAC-based mammalian artificial chromosomes. *Nat. Biotech.* **16,** 431–439.

8. Trowell, H. E., Nagy, A., Vissel, B., and Choo, K. H. A. (1993) Long-range analyses of the centromeric regions of human chromosomes 13, 14 and 21: Identification of a narrow domain containing two key centromeric DNA elements. *Hum. Mol. Genet.* **2,** 1639–1649.

9. Ikeno, M., Masumoto, H., and Okazaki, T. (1994) Distribution of CENP-B boxes reflected in CREST centromere antigenic sites on long-range α-satellite DNA arrays of human chromosome 21. *Hum. Mol. Genet.* **3,** 1245–1257.

10. Devilee, P., Cremer, T., Slagboom, P., Bakker, E., Scholl, H. P., Hager, H. D., et al. (1986) Two subsets of human alphoid repetitive DNA show distinct preferential localization in the pericentric regions of chromosomes 13, 18, and 21. *Cytogenet. Cell Genet.* **41,** 193–201.

11. Jorgensen, A. L., Bostock, C. J., and Bak, A. L. (1987) Homologous subfamilies of human alphoid repetitive DNA on different nucleolus organizing chromosomes. *Proc. Natl. Acad. Sci. USA* **84,** 1075–1079.

12. Marcais, B., Gerard, A., Bellis, M., and Roizés, G. (1991) TaqI reveals two independent alphoid polymorphisms on human chromosomes 13 and 21. *Hum. Genet.* **86,** 307–310.

13. Greig, G. M., Warburton, P. E., and Willard, H.F (1993) Organization and evolution of an alpha satellite DNA subset shared by human chromosomes 13 and 21. *J. Mol. Evol.* **37,** 164–175.

14. Lee, C., Wevrick, R., Fisher, R. B., Ferguson-Smith, M. A., and Lin, C. C. (1997) Human centromeric DNAs. *Hum. Genet.* **100,** 291–304.

15. Alexandrov, I., Kazakov, A., Tumeneva, I., Shepelev, V., and Yurov, Y. (2001) Alpha-satellite DNA of primates: old and new families. *Chromosoma* **110,** 253–266.

16. Lo, A. W., Liao, G. C., Rocchi, M., and Choo, K. H. (1999) Extreme reduction of chromosome-specific alpha-satellite array is unusually common in human chromosome 21. *Genome Res.* **9,** 895–908.

17. Marcais, B., Bellis, M., Gerard, A., Pages, M., Boublik, Y., and Roizés G. (1991) Structural organization and polymorphism of the alpha satellite DNA sequences of chromosomes 13 and 21 as revealed by pulse field gel electrophoresis. *Hum. Genet.* **86,** 311–316.

18. Mashkova, T., Oparina, N., Alexandrov, I., Zinovieva, O., Marusina, A., Yurov, Y., et al. (1998) Unequal cross-over is involved in human alpha satellite DNA rearrangements on a border of the satellite domain. *FEBS Lett.* **441,** 451–457.

19. Horvath, J. E., Viggiano, L., Loftus, B. J., Adams, M. D., Archidiacono, N., Rocchi, M., et al. (2000) Molecular structure and evolution of an alpha satellite/non-alpha satellite junction at 16p11. *Hum. Mol. Genet.* **9,** 113–123.

20. Greig, G. M., England, S. B., Bedford, H. M., and Willard, H. F. (1989) Chromosome-specific alpha satellite DNA from the centromere of human chromosome 16. *Am. J. Hum. Genet.* **45,** 862–872.

21. Alexandrov, I., Kazakov, A., Tumeneva, I., Shepelev, V., and Yurov, Y. (2001) Alpha-satellite of primates: Old and new families. *Chromosoma* **110,** 253–266.

22. Schueler, M. G., Higgins, A. W., Rudd, M. K., Gustahaw, K., and Willard H. F. (2001) Genomic and genetic definition of a functional human centromere. *Science* **294,** 109–115.

23. Guy, J., Spalluto, C., McMurray, A., Heran, T., Crosier, M., Viggiano, L., et al. (2000) Genomic sequence and transcriptional profile of the boundary between pericentromeric satellites and genes on human chromosome 10q. *Hum. Mol. Genet.* **9,** 2029–2042.

24. Horvath, J. E., Viggiano, L., Loftus, B. J., Adams, M. D., Archidiacono, N., Rocchi, M., et al. (2000) Molecular structure and evolution of an alpha satellite/non-alpha satellite junction at 16p11. *Hum. Mol. Genet.* **9,** 113–123.

25. Hattori M., Fujiyama A., Taylor T. D., Watanabe H., Yada T., Park H. S., et al. (2000) The DNA sequence of human chromosome 21. *Nature* **405,** 311–319.

26. Mashkova, T. D., Tyumeneva, I. G., Zinoveva, O. L., Romanova, L. Y., Jabs, E., and Aleksandrov, I. A. (1996) Centromeric alpha-satellite DNA at euchromatin/heterochromatin boundary of human chromosome 21. *Mol. Biol.* **30,** 617–625.

27. Laurent A. M., Puechberty, J., Prades, C., Gimenez, S., and Roizés G. (1997) Site-specific retrotransposition of L1 elements within human alphoid satellite sequences. *Genomics* **46,** 127–132.

28. Puechberty, J., Laurent, A. M., Gimenez, S., Billault, A., Brun-Laurent, M. E., Calenda, A., et al. (1999) Genetic and physical analyses of the centromeric and pericentromeric regions of human chromosome 5: Recombination across 5cen. *Genomics* **56,** 274–287.

29. Laurent, A. M., Puechberty, J., and Roizés, G. (1999) Hypothesis: For the worst and for the best, L1Hs retrotransposons actively participate in the evolution of the human centromeric alphoid sequences. *Chrom. Res.* **7,** 305–317.

30. Eichler, E. E. (2001) Recent duplication, domain accretion and the dynamic mutation of the human genome. *Trends Genet.* **17,** 661–669.

31. Horvath, J. E., Bailey, J. A., Locke, D. P., and Eichler, E. E. (2001) Lessons from the human genome: Transitions between euchromatin and heterochromatin. *Hum. Mol. Genet.* **10,** 2215–2223.

32. Brewer, C., Holloway, S., Zawalnyski, P., Schinzel, A., and FitzPatrick, D. (1999) A chromosomal duplication map of malformations: Regions of suspected haplo—and triplolethality—and tolerance of segmental aneuploidy. *Am. J. Hum. Genet.* **64,** 172–1708.

33. Hilliker, A. J. and Appels, R. (1982) Pleiotropic effects associated with the deletion of heterochromatin surrounding rDNA on the X chromosome of Drosophila. *Chromosoma* **86,** 469–490.

34. Wu, C. I., True, J., and Johnson, N. (1989) Fitness reduction associated with the deletion of a satellite DNA array. *Nature* **341,** 248–251.

35. Brun, M. E., Ruault, M., Ventura, M., Roizés, G., and De Sario, A. (2003) Juxtacentromeric region of human chromosome 21: a boundary between centromeric and heterochromatic and euchromatic chromosome arms. *Gene,* in press.

36. Gatti, M. and Pimpinelli, S. (1992) Functional elements in *Drosophila melanogaster* heterochromatin. *Annu. Rev. Genet.* **26,** 239–275.

37. Eberl, D. F., Duyf, B. J., and Hilliker, A. J. (1993) The role of heterochromatin in the expression of heterochromatic gene, the rolled focus of *Drosophila melanogaster. Genetics* **134,** 277–292.

38. Howe, M., Dimitri, P., and Wakimoto, B. T. (1995) Cis-effects of heterochromatin on heterochromatic and euchromatic gene activity in *Drosophila melanogaster. Genetics* **140,** 1033–1045.

39. Guipponi, M., Yaspo, M. L., Riesselman, L., Chen, H., De Sario, A., Roizés, G., et al. (2000) Genomic structure of a copy of the human TPTE gene which encompasses 87 kb on the short arm of chromosome 21. *Hum. Genet.* **107,** 127–131.

39a. Ruault, M., Van den Bruggen, P., Brun, M. E., Boyle, S., Roizés, G., and De Sario, A. (2002) New BAGE (*B melanoma antigens*) genes mapping to the juxtacentromeric regions of human chromosomes 13 and 21 have a cancer/testis expression profile. *Eur. J. Hum. Genet.* **10,** 833–840.

40. Therman, E., Sarto, G. E., and Patau, K. (1974) Apparently isodicentric but functionally monocentric X chromosome in man. *Am. J. Hum. Genet.* **26,** 83–92.

41. Sullivan, B. A. and Willard H. F. (1998) Stable dicentric X chromosomes with two functiona centromeres. *Nat. Genet.* **20,** 227–228.

42. Agudo, M., Abad, J. P., Molina, I., Losada, A., Ripoli, P., and Villasante A. (2000) A dicentric chromosome of *Drosophila melanogaster* showing alternate centromere inactivation. *Chromosoma* **109,** 190–196.

43. Sullivan, B. A. and Schwartz, S. (1995) Idenditification of centromeric antigens in dicentric Robertsonian translocations: CENP-C and CENP-E are necessary components of functional centromeres. *Hum. Mol. Genet.* **4,** 2189–2197.

44. Warburton, P. E. (2001) Epigenetic analysis of kinetochore assembly on variant human centromeres. *Trends Genet.* **17,** 243–247.

45. Voullaire, L. E., Slater, H. R., Petrovic, V., and Choo, K. H.A. (1993) A functional marker centromere with no detectable alpha-satellite, satellite III, or CENP-B protein: Activation of a latent centromere. *Am . J. Hum. Genet.* **52,** 1153–1163.

46. Barry A. E., Bateman M., Howman E. V., Cancilla M. R., Tainton K. M., Irvine D. V., et al. (2000) The 10q25 neocentromere and its inactive progenitor have identical primary nucleotide sequence: Further evidence for epigenetic modification. *Genome Res.* **10,** 832–838.

47. Warburton P. E., Dolled M., Mahmood R., Alonso A., Li S., Naritomi K., et al. (2000) Molecular cytogenetic analysis of eight inversion duplications of human chromosome 13q that each contain a neocentromere. *Am. J. Hum. Genet.* **66,** 1794–1806.

48. Koch, J. (2000) Neocentromeres and alpha satellite: A proposed structural code for functional human centromere DNA. *Hum. Mol. Genet.* **9,** 149–154.

49. Fitzgerald, D. J., Dryden, G. L., Bronson, E. C., Williams, J. S., and Anderson, J. N. (1994) Conserved patterns of bending in satellite and nucleosome positioning DNA. *J. Biol. Chem.* **269,** 303–314.

50. Lo, A. W. I., Craig, J. M., Saffery, R., Kalitsis, P., Irvine, D. V., Earle, E., et al. (2001) A 330 kb CENP-A binding domain and altered replication timing at a human neocentromere. *EMBO J.* **20,** 2087–2096.

51. Lo, A. W. I., Magliano, D. J., Sibson, M. C., Kalitsis, P., Craig, J. M., and Choo, K. H.A. (2001) A novel chromatin immunoprecipitation and array (CIA) analysis identifies a 460-kb CENP-A-binding neocentromere DNA. *Genome Res.* **11,** 448–457.

52. Saffery R., Irvine D. V., Griffiths B., Kalitsis P., Wordeman L., and Choo K. H. (2000) Human centromeres and neocentromeres show identical distribution patterns of >20 functionally important kinetochore-associated proteins. *Hum. Mol. Genet.* **9,** 175–185.

53. Hudson, D. F., Fowler, K. J., Earle, E., Saffery, R., Kalitsis, P., Trowell, H., et al. (1998) Centromere protein B null mice are mitotically and meiotically normal but have lower body and testis weights. *J. Cell Biol.* **141,** 309–319.

54. Williams, B. C., Murphy, T. D., Goldberg, M. L., and Karpen, G. H. (1998) Neocentromere activity of structurally acentric mini-chromosomes in *Drosophila. Nat. Genet.* **18,** 30–37.

55. Yu, H. G., Hiatt, E. N., Chan, A., Sweeney, M., and Dawe, R. K. (1997) Neocentromere-mediated chromosome movement in maize. *J. Cell Biol.* **139,** 831–840.

56. Tyler-Smith, C., Ginelli, G., Giglio, S., Floridia, G., Pandya, A., Terzoli, G., et al. (1999) Transmission of a fully functional human neocentromere through three generations. *Am. J. Hum. Genet.* **64,** 1440–1444.

57. Saffery R., Wong L. H., Irvine D. V., Bateman M. A., Griffiths B., Cutts S. M., et al. (2001) Construction of neocentromere-based human minichromosomes by telomere-associated chromosomal truncation. *Proc. Natl. Acad. Sci. USA* **98,** 5705–5710.

58. Hassold, T. and Hunt P. (2001) To err (meiotically) is human: The genesis of human aneuploidy. *Nat. Rev. Genet.* **2,** 280–291.

59. Hassold, T., Merrill, M., Adkins, K., Freeman, S., and Sherman, S. (1995) Recombination and maternal age-dependent non-disjunction: Molecular studies of trisomy 16. *Am. J. Hum. Genet.* **57,** 867–874.

60. Mahtani M. M. and Willard H. F. (1998) Physical and genetic mapping of the human X chromosome centromere: Repression of recombination. *Genome Res.* **8,** 100–110.

61. Lima-de-Faria, A. and Jaworska, H. (1968) Late DNA synthesis in heterochromatin. *Nature* **217,** 138–142.

62. Dupraw, E. J. (1968) *Cell and Molecular Biology.* Academic Publishing Co., New York, p. 892.

63. McCarroll, R. M. and Fangman, W. L. (1988) Time of replication of yeast centromeres and telomeres. *Cell* **54,** 505–513.

64. Ahmad, K. and Henikoff, S. (2001) Centromeres are specialised replication domains in heterochromatin. *J. Cell Biol.* **153,** 101–109.

65. Sullivan, B. and Karpen, G. (2001) Centromere identity in *Drosophila* is not determined in vivo by replicating timing. *J. Cell Biol.* **154,** 683–690.

66. Ten Hagen K. G., Gilbert D. M., Willard H. F., and Cohen S. N. (1990) Replication timing of DNA sequences associated with human centromeres and telomeres. *Mol. Cell Biol.* **10,** 6348–6355.

67. O'Keefe R. T., Henderson S. C., and Spector D. L. (1992) Dynamic organization of DNA replication in mammalian cell nuclei: Spatially and temporally defined replication of chromosome-specific alpha-satellite DNA sequences. *J. Cell Biol.* **116,** 1095–1110.

68. Selig, S., Ariel, M., Goitein, R., Marcus, M., and Cedar, H. (1988) Regulation of mouse satellite DNA replication time. *EMBO J.* **7,** 419–426.

69. Jasencakova, Z., Meister, A., and Schubert, I. (2001) Chromatin organisation and its relation to replication and histone acetylation during the cell cycle in barley. *Chromosoma* **110,** 83–92.

70. Shelby, R. D., Monier, K., and Sullivan, K. F. (2000) Chromatin assembly at kinetochores is uncoupled from DNA replication. *J. Cell Biol.* **151,** 1113–1118.

71. Henikoff, S., Kami, A., and Malik, H. S. (2001) The centromere paradox: stable inheritance with rapidly evolving DNA. *Science* **293,** 1098–1102.

72. Litmanovitch, T., Altaras, M. M., Dotan, A., and Avivi, L. (1998) Asynchronous replication of homologous alpha-satellite DNA loci in man is associated with nondisjunction. *Cytogenet. Cell Genet.* **81,** 26–35.

73. Archidiacano, N., Antonacci, R., Marzella, R., Finelli, P., Lonoce, A., and Rocchi, M. (1995) Comparative mapping of human alphoid sequences in great apes using fluorescence *in situ* hybridization. *Genomics* **25,** 477–484.

74. Montefalcone, G., Tempesta, S., Rocchi, M., and Archidiacano, N. (1999) Centromere repositioning. *Genome Res.* **9,** 1184–1188.

75. Ventura, M., Archidiacono, N., and Rocchi, M. (2000) Centromere emergence in evolution. *Genome Res.* **11,** 595–599.

76. Malik, H. S. and Henikoff, S. (2001) Adaptative evolution of Cid, a centromere-specific histone in Drosophila. *Genetics* **157,** 1293–1298.

77. Smith, G. P. (1976) Evolution of repeated DNA sequences by unequal crossingover. *Science* **191,** 528–535.

78. Stephan, W. (1989) Tandem-repetitive noncoding DNA: Forms and forces. *Mol. Biol. Evol.* **6,** 198–212.

79. Walsh, J. B. (1987) Persistence of tandem arrays: Implications for satellite and simple-sequence DNAs. *Genetics* **115,** 553–567.

80. Fletcher, H. L. and Rafferty, J. A. (1993) The effects of unequal sister chromatid exchange on length of arrays of repeated sequences. *J. Theor. Biol.* **164,** 507–514.

81. Marcais, B., Charlieu, J. P., Allain, B., Brun, E., Bellis, M., and Roizés, G. (1991) On the mode of evolution of alpha satellite DNA in human populations. *J. Mol. Evol.* **33,** 42–48.

82. Marcais, B., Laurent, A. M., Charlieu, J. P., and Roizés, G. (1993) Organization of the variant domains of alpha satellite DNA on human chromosome 21. *J. Mol. Evol.* **37,** 171–178.

83. Shelby, R. D., Vafa, O., and Sullivan, K. F. (1997) Assembly of CENP-A into centromeric chromatin requires a cooperative array of nucleosomal DNA contact sites. *J. Cell Biol.* **136,** 501–513.

84. Strahl, B. D. and Allis, C. D. (2000) The language of covalent histone modifications. *Nature* **403,** 41–45.

85. Junewein, T. (2000) Re-SET-ting heterochromatin by histone methyltransferases. *Trends Cell Biol.* **11,** 266–273.

86. Razin, A. and Kafri, T. (1994) DNA methylation from embryo to adult. *Prog. Nucleic Acids Res. Mol. Biol.* **48,** 53–81.

87. Miniou, P., Jeanpierre, M., Bourc'his, D., Coutinho Barbosa, A. C., Blanquet, V., and Viegas-Péquignot, E. (1997) Alpha-satellite DNA methylation in normal individuals and in ICF patients: Heterogeneous methylation of constitutive heterochromatin in adult and fetal tissues. *Hum. Genet.* **99,** 738–745.

88. Xu, G-L., Bestor, T. H., Bourc'his, D., Hsieh, C-L., Tommerup, N., Bugge, M., et al. (1999) Chromosome instability and immunodeficiency syndrome caused by mutations in a DNA methyltransferase gene. *Nature* **402,** 187–191.

89. Okano, M., Bell, D. W., Haber, A., and Li, E. (1999) DNA methyltransferases Dnmt3a and Dnmt3b are essential for de novo methylation and mammalian development. *Cell* **99,** 247–257.

90. Bernardi, G. (1993) The isochore organization of the human genome and its evolutionary history-a review. *Gene* **135,** 57–66.

91. International Human Genome Consortium. (2001) Initial sequencing and analysis of the human genome, *Nature,* **409,** 860–921.

6

Identification of Start Sites of Bi-Directional DNA Synthesis at Eukaryotic DNA Replication Origins

Alessandro Vindigni, Gulnara Abdurashidova, and Arturo Falaschi

1. Introduction

DNA replication is the key process for the conservation and transmission of genetic information in all living organisms. The bi-directional DNA replication does not start randomly along the genome but rather at precise chromosomal regions, defined as origins of DNA replication where, after specific protein–DNA interactions, regulation of the process occurs.

Studies in yeast have shown that origins of replication are characterized by a conserved consensus sequence of 11 bp called autonomously replicating sequences (ARS). However, similar sequences have not been found so far in higher eukaryotic organisms. Up to now, very little has been discovered about the over 30,000 DNA replication origins that are present in the chromosomes of human cells. In fact, only five origins have been so far identified, and a common consensus sequence, like in the case of ARS in yeast, has not been found. In addition, very little is known about the pro-

From: *Methods in Molecular Biology, Vol. 240:*
Mammalian Artificial Chromosomes: Methods and Protocols
Edited by: V. Sgaramella and S. Eridani © Humana Press Inc., Totowa, NJ

tein complex that binds to these origins, and the detailed mechanism for origin activation remains to be described. For these reason, the definition of a strategy for mapping new human DNA replication origins with nucleotide resolution is of fundamental importance. In the previous years we have identified a replication origin on the G band p13.3 of chromosome 19, located within 474 bp between the site of transcription end of the gene for lamin B2 and the nontranscribed spacer preceding another, still uncharacterized gene, called ppv1 *(1)*. The subsequent study of the accessibility in vivo of the origin area to DNA-damaging agents revealed a prominent 115-bp protection during the G_1 phase (as evidenced by DMS- or DNaseI-induced fragmentation followed by ligation-mediated polymerase chain reaction [PCR] analysis of the DNA), which shrinks to a 70-bp protection at the beginning of S and disappears completely in mitosis (as well as in G_0 cells). This behavior is reminiscent of the variations occurring at the ARS1 origins in *Saccharomyces cerevisiae* that are believed to correspond to the formation of a prereplicative complex in G_1 and to its reduction to a shorter postreplicative complex immediately after origin activation in S. Recently, thanks to the strategy described in this chapter, we have been able to show that initiation of replicon synthesis occurs within the area protected in G_1- and S-phases, at two precise nucleotides overlapping by four base pairs on the two helices *(2)*. All the other human origins were identified with a resolution greater than 500 bp. Therefore, the use of our procedure for the investigation and fine mapping of other human DNA replication origins is highly demanded and will provide fundamental information for a better understanding of the mechanism of DNA replication in eukaryotic cells. In the following pages, we summarize the protocols for the proper purification of newly replicated DNA and for the precise definition of the start site of bidirectional synthesis.

2. Materials

2.1. Isolation and Analysis of Newly Replicated DNA

1. Human epithelial HeLa cells.
2. Dulbecco's modified Eagle medium (Gibco BRL, Invitrogen).

3. Fetal calf serum (GibcoBRL, Invitrogen).
4. Trypsin (New England Biolabs).
5. Proteinase K (Boehringer Mannheim).
6. RNAse inhibitor (Boehringer Mannheim).
7. *S. cerevisiae* cells strain SP1.
8. Beckmann L8-M ultracentrifuge and SW28 rotor.

2.1.1. Preparation of Size-Fractioned DNA

Total human genomic DNA can be extracted from nuclei isolated from asynchronously growing cells as follows. Human epithelial HeLa cells are grown in Dulbecco's modified Eagle medium supplemented with 10% fetal calf serum and antibiotics. Semiconfluent cultures are detached by treatment with trypsin, washed first in complete medium and then in cold phosphate-buffered saline. Nuclei are isolated by resuspending cells in RSB buffer (10 mM Tris-HCl pH8; 10 mM NaCl; 3 mM MgCl$_2$) at a concentration of 2.5×10^7 cells/mL and incubating the sample in ice for 5 min. An equal volume of 0.2% NP-40 in RSB buffer is added followed by incubation in ice for additional 10 min. The nuclei so obtained are pelleted by centrifugation and treated by lysis buffer (200 mM NaCl; 10 mM Tris-HCl, pH 8; 25 mM ethylenediamine tetraacetic acid (EDTA); 1% sodium dodecyl sulfate (SDS); and 600 µg/mL proteinase K) overnight at 37°C. The sample is then extracted twice with phenol:chloroform:isoamyl alcohol (25:24:1) and once with chloroform:isoamyl alcohol. After extraction, an equal volume of isopropanol is added to the aqueous phase and the precipitate is collected by centrifugation for 30 min at 4°C. DNA is then resuspended in TE buffer (10 mM Tris-HCl, pH 8; 1 mM EDTA) in the presence of 1 U/mL of RNase inhibitors (Boehringer Mannheim) and denatured by heating for 5 min at 95°C.

Yeast genomic DNA is isolated from asynchronous *S. cerevisiae* cells strain SP1 grown at 2×10^7 cell/mL cell density. Cells are harvested by centrifugation, washed in sorbitol buffer (1 M sorbitol, 100 mM EDTA), and resuspended in spheroplast buffer (1 M sorbitol, 100 mM EDTA, 100 mM β-mercaptoethanol, 3 U/µL zymolase).

Spheroplasts are resuspended in lysis buffer (50 mM Tris-HCl, pH 8.0; 50 mM EDTA; 1% SDS; 1 mg/mL of proteinase K). After lysis and debris precipitation using 3 M potassium acetate, nucleic acids are extracted twice with phenol:chloroform:isoamyl alcohol, precipitated with isopropanol, and resuspended in TE buffer. To obtain nascent DNA from total human and yeast genomic DNAs, samples of 150 μg and 12 μg, respectively, are denatured 5 min at 95°C and fractionated on a 5 to 30% (w/v) linear sucrose gradient (35 mL) by centrifugation for 17 h at 20°C in a Beckman SW28 rotor at 90,000g.

In a parallel gradient, double-stranded DNA size markers are fractionated. Thirty-four 1-mL fractions are collected for both gradients and aliquots of the marker fractions are analyzed on agarose gels. Using the separation pattern of the marker DNA as sedimentation velocity reference, fractions containing ssDNA ranging in size between 800 and 1200 bp are pooled and the DNA is recovered by ethanol precipitation and resuspended in 20 μL of TE buffer.

2.1.2. Enrichment of RNA-Primed Nascent DNA by Treatment With l-Exonuclease

2.1.2.1. MATERIALS

1. Polynucleotide kinase (New England Biolabs).
2. ATP (Amersham).
3. λ-exonuclease (Gibco BRL).
4. RNAse A, RNAse T1, and RNAse T2 (Boehringer Mannheim).
5. Proteinase K (Boehringer Mannheim).

2.1.2.2. METHODS

To render all DNA molecules in the purified nascent DNA samples available for λ-exonuclease digestion, samples (0.2 to 2 μg) are treated with polynucleotide kinase (New England Biolabs) to phosphorylate 5 hydroxyl ends. The reaction is conducted in a total volume of 50 μL with 2.5 μL of enzyme in the presence of 1 mM ATP and using the condition described by Gerbi and collaborators *(3)* . The same reaction is also conducted on total DNA samples

(40 μg) extracted from cells synchronized at different points in the cell cycle. The phosphorylated DNA is resuspended in water and with 2 μL of λ-exonuclease (Gibco BRL) in 67 mM glycine-KOH, pH 8.8; 2.5 mM MgCl$_2$; and 50 μg/mL bovine serum albumin in a final volume of 20 μL at 37°C overnight. The enzyme is then heat inactivated by 10-min incubation at 75°C and the samples are extracted once with phenol:chloroform:isoamyl alcohol and once with chloroform. Ethanol precipitation is performed in dry ice for 20 min in the presence of 0.3 M sodium acetate, pH 7.2, and 2 μL of glycogen (20 μg/μL) as a carrier. The DNA pellet is resuspended in TE. Covalently bound RNA is degraded by incubation with RNAse A (Boheringer Mannheim, final concentration: 100 μg/mL), RNAse T1 (Boheringer Mannheim, final concentration: 1 U/μL), and RNAse T2 (Gibco BRL, final concentration: 0.6 U/μL) overnight at 37°C, followed by protein digestion in the same buffer supplemented with 0.1% SDS and 600 μg/mL proteinase K. DNA is then purified by phenol:chloroform extraction and ethanol precipitation as described above. Alternatively, RNA is degraded by heating samples for 5 min at 95°C in 0.1 M NaOH. The solution is neutralized with HCl and the DNA was recovered by ethanol precipitation. After RNA removal, samples are resuspended in water and subjected to a second kinase reaction, to phosphorylate the 5' hydroxyl DNA ends exposed on nascent DNA after RNA primer digestion. The reaction is conducted in the conditions defined previously. The DNA, purified and recovered by precipitation as described previously, is finally resuspended in 20 μL TE buffer. Aliquots of the samples before and after λ-exonuclease treatment are now ready analyzed by competitive PCR or ligation-mediated PCR (LM-PCR).

2.1.3. Competitive PCR Analysis of Nascent DNA

2.3.1.1. MATERIALS

1. Oligonucleotide primers.
2. PCR machines.
3. Taq polymerase (Boehringer Mannheim).
4. Nocodazole (Sigma).
5. λ-exonuclease (New England Biolabs).

2.3.1.2. METHODS

The quantitation of the abundance of two different sequences in the lamin B2 gene region is performed by competitive PCR as already described *(4,5)*. The sequences chosen for amplification are defined as B48 and B13 (**Fig. 1**); the former corresponds to the lamin B2 ori (within 474 nt), whereas B13 is displaced by 5 kb. A single competitor molecule is used for both markers. The core of the molecule is a 110-bp region derived from the β-globin gene harboring a 20-bp insertion. The forward and reverse primers spanning the lamin B2 origin area (for details, *see* **ref. 6**) arranged in a head-to-tail fashion are joined to the core molecule by PCR amplifications using chimerical primers. This approach allows the use of a single competitor for the quantification of the relative abundance of different PCR markers. The competitor is quantified in competitive PCR experiments against a known amount of plasmid molecules harboring the lamin B2 origin area. The PCR cycle profile is as follows: denaturation at 94°C, annealing at 56°C, and extension at 72°C; time for each step is 30 s; 35 cycles are performed with 1 U of Taq polymerase (Boehringer Mannheim) in the conditions recommended by the manufacturer. Competitive PCR experiments are conducted challenging the same amount of size-selected DNA from asynchronous cells (before and after the λ-exonuclease treatment) and total DNA from synchronized cells (after the λ-exonuclease treatment) with 10-fold serial dilutions of the competitor. Further details about the procedure for competitive PCR mapping are contained in **ref. 5**.

Figure 1 reports the results of the analysis of enrichment in lamin B2 origin sequences in the different DNA preparations. In **Fig. 1A**, schematic representation of the human lamin B2 genomic region containing the origin of DNA replication that we are exploring. The bubble structure marks the position of the ori, the arrows show the locations of the lamin B2 and ppv1 transcripts. The position of the regions amplified by competitive PCR is shown in black boxes. In **Fig. 1B**, competitive PCR analysis of size fractionated nascent DNA isolated from asynchronously growing HeLa cells or from human normal G0 lymphocytes. DNA samples are analyzed for their

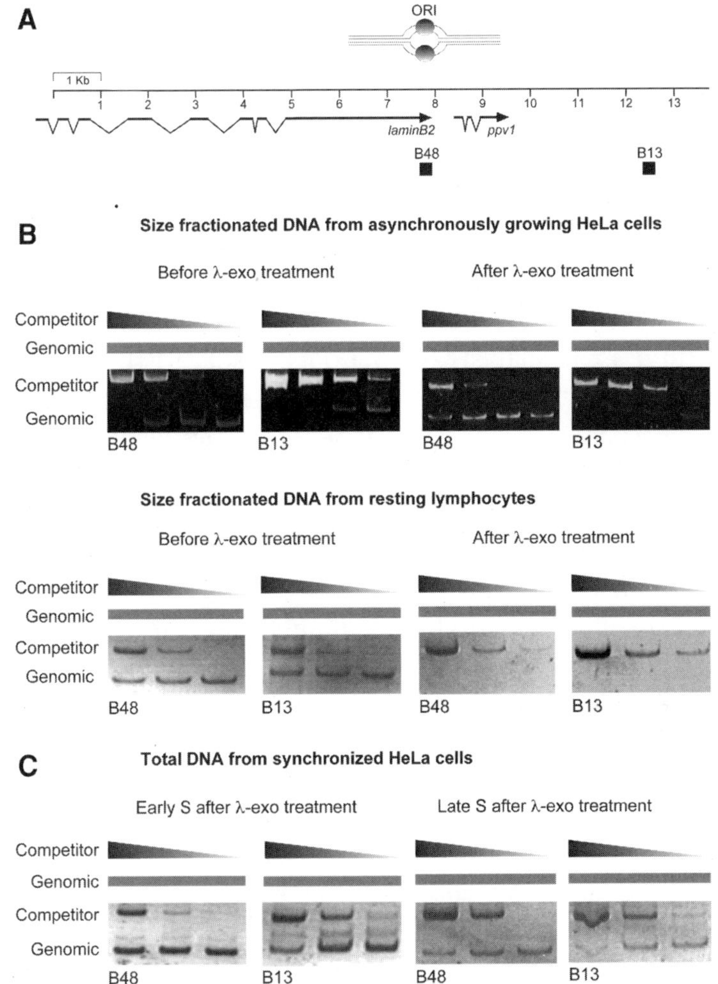

Fig. 1. Analysis of newly replicated DNA by competitive PCR. (**A–C**) *See* opposite page for details.

abundance in origin (B48) or control (B13) sequences before or after λ-exonuclease treatment. For each region a fixed amount of DNA is coamplified with an amount of competitor molecules decreasing by 10-fold dilution. Before each nascent DNA quantification, it is verified that B48 and B13 regions are amplified with the same efficiency on total genomic DNA. The number of competitor molecules per PCR ranges from 104 to 101 molecules. As the results show, the nascent DNA isolated by size fractionation from asynchronously growing cells displays a 10-fold of enrichment of ori-containing sequences with respect to the control one (**Fig. 1B**). By hydrolysis with λ-exonuclease we obtained a further enrichment reaching thus a greater than 50-fold ratio of origin over control sequences. The same experiment performed on human quiescent lymphocytes does not show any enrichment of B48 over B13 nor does it show the presence of any DNA after λ-exonuclease treatment. In **Fig. 1C**, competitive PCR analysis on total DNA isolated from HeLa cells synchronized at the beginning and at the end of S-phase. For synchronization experiments, cells at 50–60% confluence are accumulated in M-phase by incubation for 10 h with 50 ng/mL nocodazole (Sigma). The cells are then collected by shake-off, washed with complete medium, and replated. Three hours later, aphidicolin is added to cell monolayer at the concentration of 5 μg/mL and the culture is incubated for 20 h to accumulate cells at the G1/S border. The cells are then released into S-phase in complete medium and are collected at different time points in S-phase. In good agreement with the finding that lamin B2 fires within the first minute of S, we observe a fivefold enrichment of B48 over B13 region on total DNA isolated from cells collected in early S and treated with λ-exonuclease (**Fig. 1C**). Instead, in late S, when the lamin B2 replicon is probably entirely duplicated, no enrichment of B48 over B13 is observed (**Fig. 1C**).

2.2. Assessment of LM-PCR as a Tool to Map the Start Sites of Nascent DNA Synthesis

2.2.1. LM-PCR Procedure

2.2.1.1. Materials

1. Vent polymerase (New England Biolabs).
2. dNTPS (Amersham).
3. DNA ligase (New England Biolabs).
4. Oligonucleotide primers and PRC machines.
5. DNA sequencing gel equipment.

2.2.1.2. Methods

The LM-PCR experiments are performed essentially according to the protocol of Quivy and Becker *(6)*. For each amplification, a set of three primers is used (*see* **Figs. 2–5**). The DNA samples are denatured 5 min at 95°C and annealed to 0.1 pmol of primer 1 in each set for 30 min in 10 mL of 1 Vent polymerase buffer (New England Biolabs). After incubation, second-strand DNA synthesis is allowed to occur by adding the four dNTPS (final concentration 0.5 mM) and Vent exo- (1 U, New England Biolabs) in a final volume of 20 µL. Extension is performed by incubation at the annealing temperature of primer 1 in each set for 5 min, followed by a 10 s incubation at 65°C, 10 s at 70°C, and 10 min at 76°C. The elongation product is then ligated to the double-stranded asymmetric linker already described *(6)*. For ligation, 20 µL of the extension reaction are mixed with 5 µL ligase buffer (New England Biolabs), 5 µL annealed asymmetric linker (200 pmol/µL), and 19 µL of 40% polyethylene glycol 8000. One microliter of DNA ligase (400 U, New England Biolabs) is then added and the reaction is allowed to occur overnight at 16°C. The ligation reaction is mixed to 150 µL TE, extracted with phenol:chloroform, and precipitated with ethanol in the presence of 2 µL glycogen (20 µg/µL) and 0.3 M ammonium acetate. The DNA pellet is washed in 70% ethanol, dried, and resuspended in 20 µL water. The ligated product is subjected to PCR amplification in 1X Vent buffer, 4 mM $MgSO_4$, 0.2 mM dNTPS using 10 pmol of primer 2 in each set, 10 pmol of long linker primer for LM-PCR *(6)*, and 1 U of Vent polymerase in a total volume of 50 µL. The PCR conditions are as follows: pre-PCR denaturation: 3 min, 95°C; denaturation: 1 min at 95°C; annealing: 2 min at the temperature indicated below for each primer 2; extension: 3 min at 76°C plus 3 s added at any further cycle. After 18 cycles of amplification, 20-µL the reaction is subjected to radioactive extension in the presence of 0.8 pmol of γ^{32}P-labeled primer 3 in 1X Vent buffer, 2.6 mM $MgSO_4$, 0.2 mM dNTPS, and 0.5 U of Vent exo- in a final volume of 30 µL. The conditions of the radioactive extension are as follows: pre-PCR denaturation: 3 min at 95°C; denaturation: 1 min at 95°C; annealing: 2 min at the temperature indicated for each

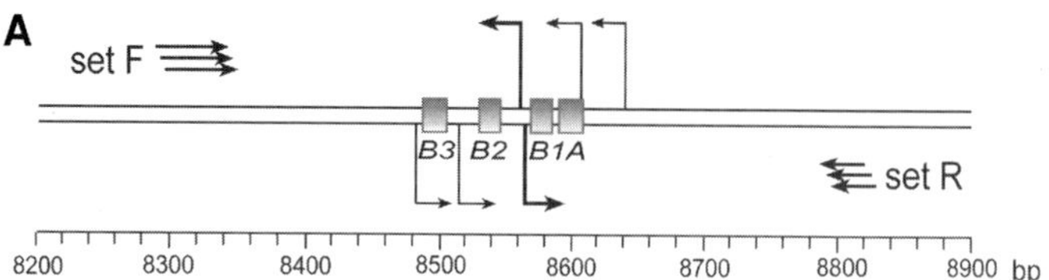

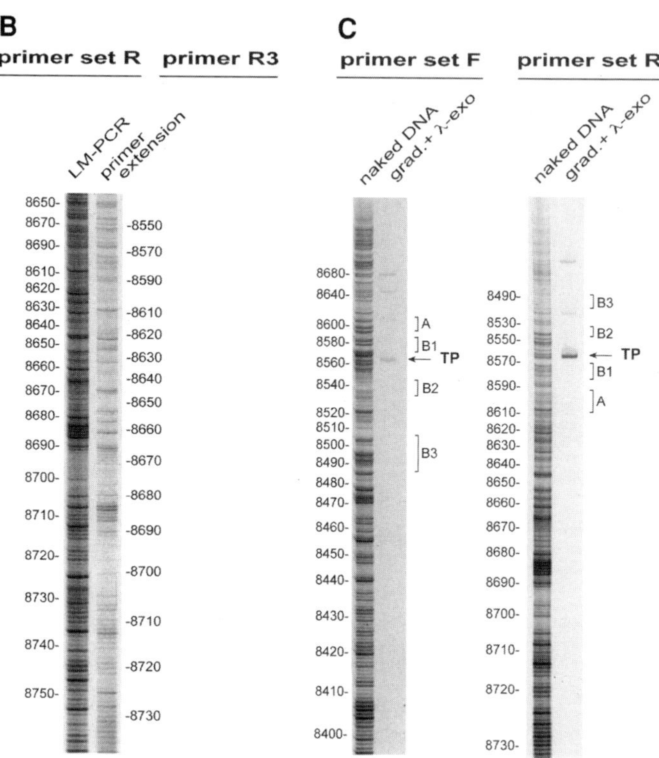

Fig. 2. Identification of precise start sites of nascent DNA synthesis within the protected area. (**A**) Yeast ARS1 region. Positions of sequence elements A, B1, B2, and B3 are marked by boxes. Localization and orientation of primer sets Fy and Ry used for LM-PCR are shown (**Note 1**). (**B**) Comparison of the size distribution of amplified and not amplified molecules. Ten micrograms of yeast genomic DNA were treated with dimethyl sulfate (final concentration 0.25%) for 2 min at room temperature, and 0.01 µg of total genomic DNA was analyzed by LM-PCR with primer set R (left lane); 2 µg of the same DNA was subjected to primer extension with primer set Ry 3 (right lane). As the figure shows, the

primer 3; extension: 7 min at 95°C; five cycles of extension are performed. The extension reaction is blocked by adding EDTA (final concentration 40 m*M*). The DNA is purified by phenol: chloroform extraction and precipitated with ethanol in the presence of 0.3 *M* sodium acetate, washed with 70% ethanol dried briefly, and resuspended in loading buffer (95% formamide, 100 m*M* EDTA, 0.5% bromophenol blue, 2.5% xylene cyanol). The samples are denatured 3 min at 95°C and loaded on a 8% sequencing acrylamide gel. An example of the results obtained by LM-PCR on the yeast ARS1 region and lamin B2 replication origin is shown in **Figs. 2–5**.

3. Notes

1. The sequences of the primer sets and of the respective annealing temperatures used in LM-PCR analysis of yeast nascent DNA are as follows. Primer set F: Fy1 8291-8315 (60°C), Fy2 8319–8345 (68°C), Fy3 8319–8349 (72°C). Primer set R: Ry1 8826–8801 (60°C), Ry2 8815–8788 (68°C), Ry3 8813–8784 (72°C). Three of the four primer sets used for the human nascent upper strand analysis (A, C, and E) and for the nascent lower strand analysis (set B and D) have been previously described (*8*). The sequences of the other primer sets used are as follows: F1 3712–3742 (60°C), F2 3693–3718 (68°C), F3 3682–3712 (72°C); G1 3573–359460°C), G2 3584–3609 (68°C), G3 3593–3619 (70°C); H1 3837–3858 (60°C), H2 3842–3871 (68°C), H3 3867–3900 (70°C). The numbering refers to the file humlambbb of GenBank (accession number M94363).

Fig. 2. (*continued from opposite page*): pattern of single-strand extended fragments and of the LM-PCR amplified ones are exactly superimposable, allowing for the difference in size given by the ligated linker. (C) LM-PCR analysis of 0.01 μg of total genomic DNA or of approx 10^4 molecules of size-fractionated, 1-exonuclease-digested nascent DNA isolated from asynchronously growing yeast cells. Positions of the transition points (TP) from continuous to discontinuous DNA synthesis are shown to correspond to the published data (*7*).

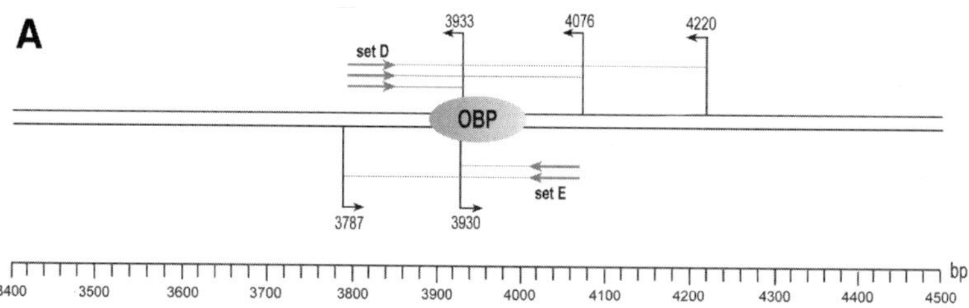

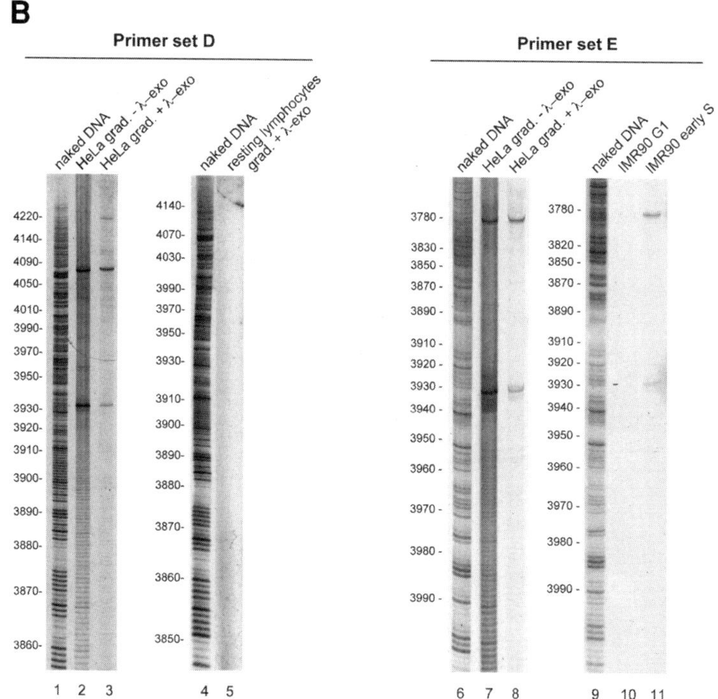

Fig. 3. Indentification of precise start sites of nascent DNA synthesis around the lamin B2 ori. (**A**) Localization and orientation of the primer sets D and E in the analyzed 1-kb region (**Note 1**). Position of the first detectable start site of leading-strand synthesis is indicated by thick arrows, and start sites of nascent DNA synthesis located more upstream are indicated by thinner arrows, all pointing in the direction of synthesis. (**B**) LM-PCR analysis on lower and upper strand, respectively, of size-fractionated newly replicated DNA from asynchronously growing HeLa cells or total DNA from synchronized IMR-90 cells subjected or not subjected to λ-exonuclease treatment as indicated.

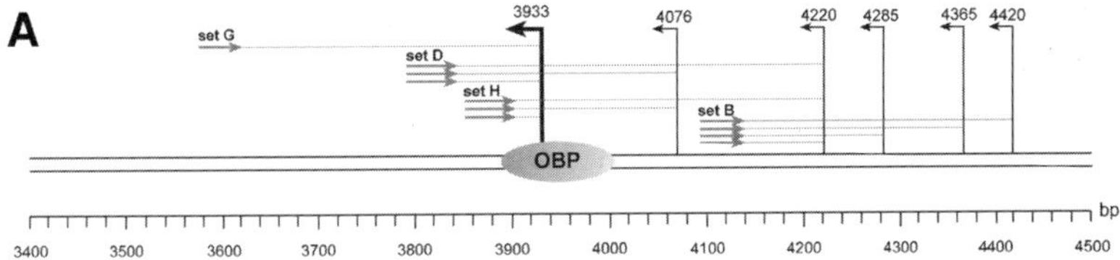

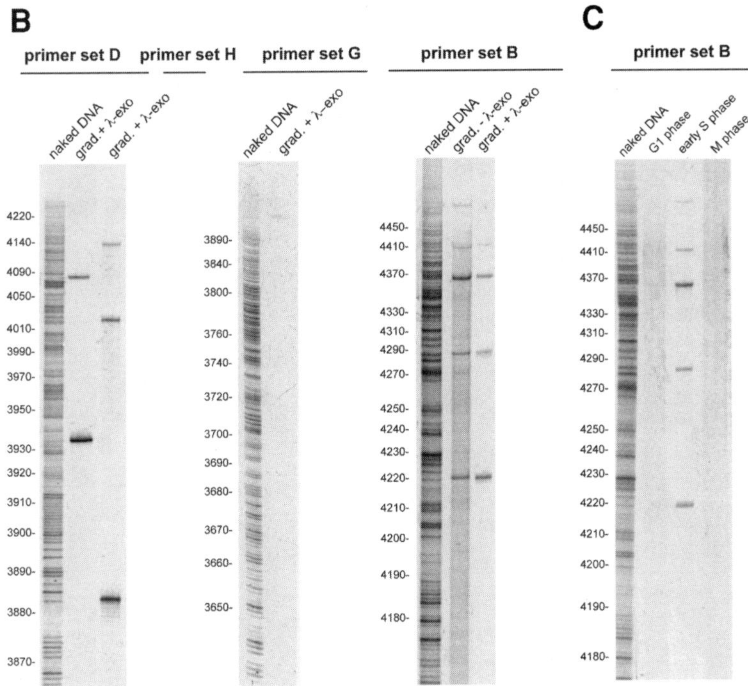

Fig. 4. Distribution of start sites of nascent DNA synthesis around the lamin B2 ori, upper nascent strand. (**A**) Localization and orientation of the primer sets D, H, G, and B used to identify the position of RNA-DNA junctions on the upper nascent strand from nucleotide 3400 to 4500 (**Note 1**). Position of the first detectable start site of leading-strand synthesis is indicated by thick arrows, and start sites of nascent DNA synthesis located more upstream are indicated by thinner arrows, all pointing in the direction of synthesis. (**B**) LM-PCR analysis of size-fractionated newly replicated DNA from asynchronously growing HeLa cells subjected to λ-exonuclease treatment. (**C**) LM-PCR analysis of total genomic DNA isolated from cells synchronized at the indicated phase of the cell cycle.

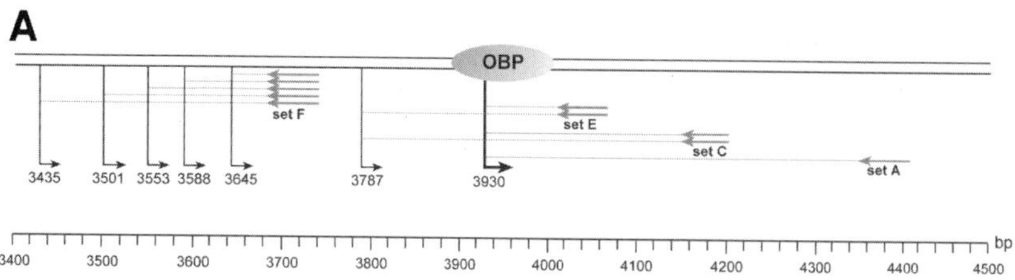

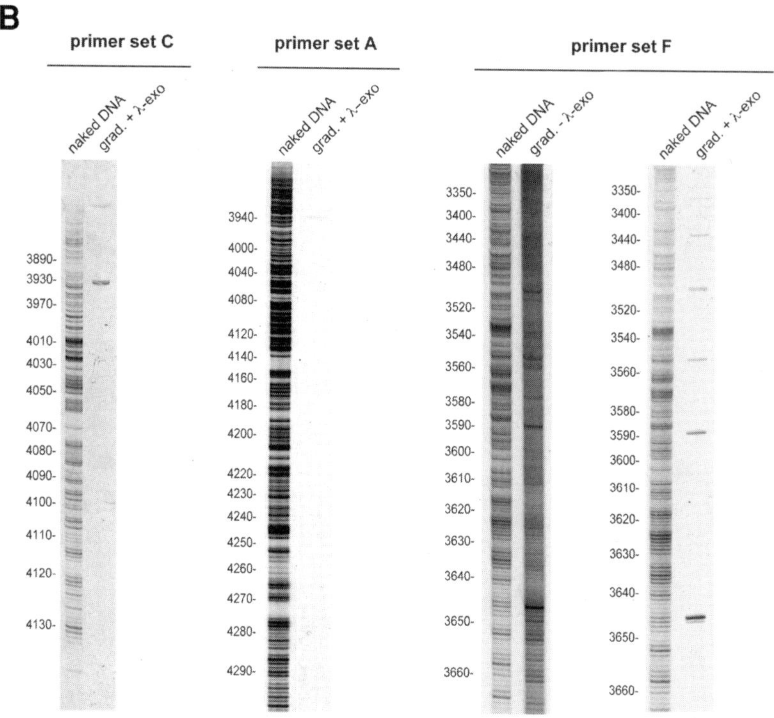

Fig. 5. Distribution of start sites of nascent DNA synthesis around the lamin B2 ori, lower nascent strand. (**A**) Localization and orientation of the primer sets A, C, E, and F used to identify the position of RNA-DNA junctions on the upper nascent strand from nucleotide 4350 to 3400 (**Note 1**). Position of the first detectable start site of leading-strand synthesis is indicated by thick arrows, and start sites of nascent DNA synthesis located more upstream are indicated by thinner arrows, all pointing in the direction of synthesis. (**B**) LM-PCR analysis of size-fractionated newly replicated DNA from asynchronously growing HeLa cells subjected or not subjected to λ-exonuclease treatment as indicated.

4. Conclusions

Our procedure allows the definition with nucleotide resolution of the start site of bi-directional replication in DNA replication origins. The results obtained on the lamin B2 DNA replication origin are shown in the **Figs. 3–5**. The search for the possible start sites at the lamin B2 ori commenced with primer sets D and E (**Fig. 3**) firing from either side of the protected region toward each other. With primer set D three discrete stop sites of second-strand synthesis (and hence of RNA-DNA junction) are visible: the first at nucleotide 3933, the second at nucleotide 4076, and the third at a less precise position; these data are much more evident after 1-exonuclease treatment. The analysis with primer set E gives specular results: two clear sites of RNA-DNA junction are visible, one at nucleotide 3930 and the other at nucleotide 3787. The same experiment on size-fractionated DNA from quiescent human lymphocytes gave no evidence of any stop site for primer set D. Furthermore, with DNA extracted from IMR-90 cells synchronized in G_1-phase and at the beginning of S, analysed with primer set E, no stop sites are visible in G_1 cells, whereas those that have just entered S show precise stop sites in exactly the same positions as the asynchronously growing cells (**Fig. 3**). Thus, nucleotides 3933 and 3930, located within the protected area, might correspond to the starts of the leading strands moving leftward and rightward, respectively, whereas the RNA-DNA junctions observed about 140 nucleotides upstream of both could derive from ligation of the first Okazaki fragment to the respective leading-strand start.

To further strengthen these conclusions, we analyzed upper nascent DNA strands with primer sets H, D, G, and B (**Figs. 4** and **5**). The upper nascent DNA strands was analyzed with primer sets H, G, and B (*see* **Fig. 4B,C**). In agreement with the data obtained with primer D, the two primers H and G indicated nucleotide 3933 as the likely start site of a leading strand. This conclusion was confirmed with primer set B, located approx 200 nucleotides to the right of nucleotide 3933; several stop sites are visible at the indicated positions. These data can be interpreted as deriving from the linking to

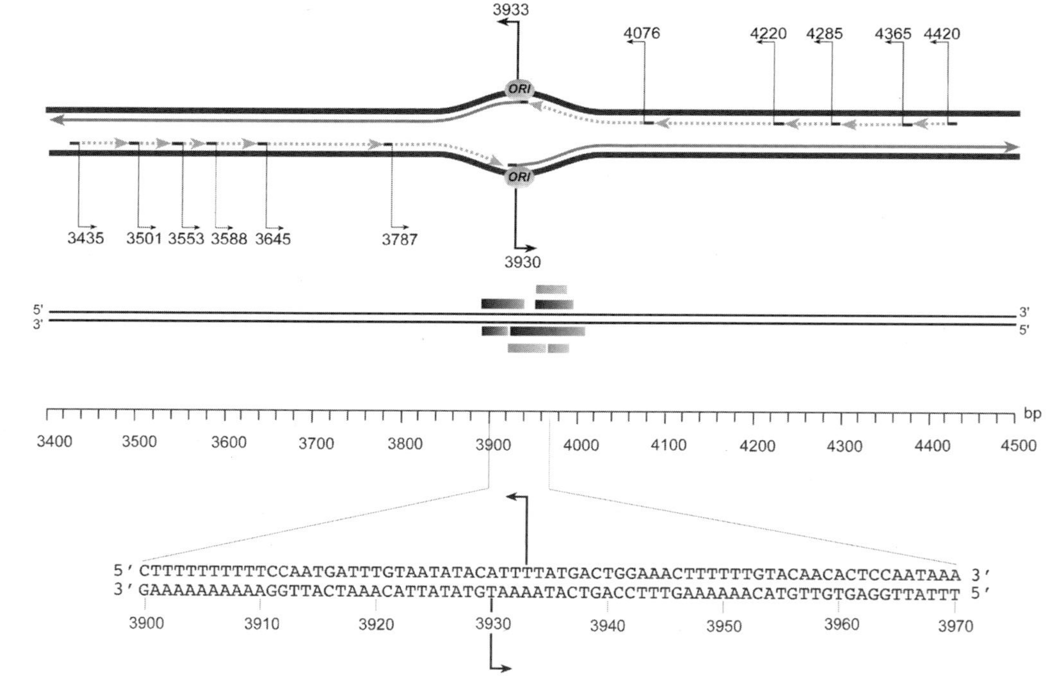

Fig. 6. Map of start sites of bi-directional DNA synthesis at the human lamin B2 ori. Transition points (TP) from continuous to discontinuous DNA synthesis on the two strands are shown by thick arrows; the 5' ends of Okazaki fragments are indicated by thin arrows. Positions of the transcripts around ori area are shown. Precise positions of the protein–DNA interactions occurring in G_1 and S phase are shown by gray and black boxes, respectively.

the initiated ori of four subsequent Okazaki fragments having lengths of 143, 144, 65, and 80 nucleotides from the first to the fourth, respectively. The lower nascent DNA strands was instead analyzed with primer sets C, A (displaced to the right of nucleotide 3930), and F (displaced to its left; **Fig. 5A,C**). With primer sets C and A, we obtained again a clear indication that the first stop site corresponds to nucleotide 3930. With primer set A the long, >450-nucleotide region free of stop sites indicates that we are dealing with an area of continuous leading-strand synthesis and that nucleotide 3930 is the start site of the rightward moving leading strand. Conversely, with primer set F a number of start sites can be located precisely at the indicated positions, possibly corresponding to ligation of the first Okazaki fragments, having lengths (in the order of fork movement) of 143, 142, 67, 35, and 52, plus two more of approx 65 nucleotides.

The conclusion of our studies is that the initiation of replicon synthesis on the lamin B2 origin occurs in a well-defined area overlapping by 4 bp (**Fig. 6**). In this way, the start site of a human replication origin was characterized for the first time with nucleotide resolution. The extension of this approach to other already known metazoan origins will be essential for a better comprehension of the mechanism of DNA replication and, more in particular, the sequence alignment of the different origins mapped with our procedure will help the discovery of possible conserved sequence motifs essential for eukaryotic DNA replication.

References

1. Abdurashidova, G., Riva, S., Biamonti, G., Giacca, M., and Falaschi, A. (1998), Cell cycle modulation of protein-DNA interactions at a human replication origin. *EMBO J.* **17,** 2961–2969.
2. Abdurashidova, G., Deganuto, M., Klima, R., Riva, S., Biamonti, G., Giacca, M., and Falaschi, A. (2000) Start sites of bidirectional DNA synthesis at the human lamin B2 origin. *Science* **287,** 2023–2026.
3. Bielinsky, A. K., and Gerbi, S. A. (1998) Discrete start sites for DNA synthesis in the yeast ARS1 origin. *Science* **279,** 95–98.

4. Giacca, M., Pelizon, C., and Falaschi, A. (1997) Mapping replication origins by quantifying relative abundance of nascent DNA strands using competitive polymerase chain reaction. *Methods* **13,** 301–312.

5. Giacca, M., Zentilin, L., Norio, P., Diviacco, S., Dimitrova, D., Contreas, G., et al. (1994) Fine mapping of a replication origin of human DNA. *Proc. Natl Acad. Sci. USA* **91,** 7119–7123.

6. Quivy, J. P. and Becker, P. B. (1993) An improved protocol for genomic sequencing and footprinting by ligation-mediated PCR. *Nucleic Acids Res.* **21,** 2779–2781.

7. Hay, R. T. and DePamphilis, M. L. (1982) Initiation of SV40 DNA replication in vivo: Location and structure of 5 ends of DNA synthesized in the ori region. *Cell* **28,** 767–779.

7

Telomere Length Analysis and In Vitro Telomerase Assay

Fiorentina Ascenzioni, Pier Assunta Fradiani, and Pierluigi Donini

1. Introduction

Barring exceptional instances, the DNA contained in eukaryotic chromosomes is linear. Linearity of the chromosomal DNA and the compartmentalized architecture of the eukaryotic cell are the two principle features that distinguish the prokaryotes from the eukaryotes and that have facilitated the evolution of totally different strategies for interaction with other species and the environment. Major changes in strategy that were made possible by linearity of chromosomes were an enormous increase in the information content of the genome and the development of sexuality as a means for efficient exchange of genetic information *(1)*. But linear chromosomes have ends, and the presence of ends produced two major biological problems. One problem was identified early by Muller *(2,3)* and by McClintock *(4)*, as the requirement to protect natural ends of chromosomes from fusion and recombination with other chromosomes and from exonucleolytic erosion. It was thus recognized that the ends of chromosomes must have specialized structural and functional features required for chromosome stability. We know today

From: *Methods in Molecular Biology, Vol. 240:*
Mammalian Artificial Chromosomes: Methods and Protocols
Edited by: V. Sgaramella and S. Eridani © Humana Press Inc., Totowa, NJ

that another potential cause of chromosomal instability that must be dealt with is that unless it is protected, a DNA end will be recognized as damaged DNA that the cell will attempt to heal with ensuing loss of chromosome integrity and cell viability. The second problem was identified more recently, when it became evident that all known DNA polymerases are unable to start synthesis *de novo* but can only add a nucleotide onto an already existing 3'-OH group, normally provided by an RNA primer. Such a primer positioned at the very end of the chromosomal DNA would be degraded, shortening the lagging DNA strand. Both these sets of problems have found a solution through the development of a unique chromosomal organelle, the *telomere*, comprised of particular telomeric DNA sequences complexed with particular *non-histone* proteins, that are organized at least in part in a *non-nucleosomal* heterochromatic form of chromatin *(5)*, defined as the *telosome*.

Telomeric DNA must be extended by a specialized mechanism to counteract replicative erosion. Eukaryotes accomplish this by making use of three distinct but interrelated mechanisms, each of which requires the presence of telomeric DNA with specific properties. In most eukaryotic organisms telomeric DNA consists of tandem arrays of short (5–8 bp) repeats of characteristic telomeric sequences with the strand replicated by leading strand synthesis being rich in G clusters. The first telomeric sequences to be determined were T_2G_4 in *Tetrahymena* *(6)* and T_4G_4 in the hypotrichous ciliates *(7)*. The most common sequences found are T_2AG_3 in the vertebrates and in several simple eukaryotes and T_3AG_3 in most plants. The telomeric DNA repeats of other eukaryotes are similar in length and sequence to the above, with the exception of a number of yeasts where the repeats are less G rich and have lengths ranging from 8 to 26. There is an enormous spread in the size of the telomeric DNA in different species: it ranges from 36 bp in some hypotrichous ciliates to 150 kb in tobacco. Telomeric DNA belonging to this class is double stranded over most of its length, but terminates in a 3' extension of the G-rich strand. The repeats are extended by reverse transcription, making use of a riboenzyme, telomerase. Telomerase acts by reverse transcribing part of the its RNA subunit into DNA, thereby synthesizing the repeats that are added to the G-rich strand of the telomeric DNA. In some cases where proliferation occurs in cells in which telomerase is inactive or missing, telomere elongation can be conducted by recombination. In fact this back-up mechanism is dependent on RAD52 in budding yeast *(8,9)* and on RAD 51 and RAD52 *(10)* in mammalian cells. A second class of telomeric DNA has been studied most extensively in the dipteran *Chironomus*. Here, the telomeric DNA ends in 200 kb of 340 bp tandem repeats *(11)*, and it has been suggested that such long repeats are extended by recombination. Recombination has also been suggested as the mechanism that extends the telomeres in *Anopheles gambiae (12)*. A third class of telomeres is found in *Drosophila melanogaster*. In this organism, that lacks telomerase, telomeric DNA consists of two families of non-long terminal repeat (LTR) retrotransposable elements, Het-A and TART, and telomere extension is achieved by specific transposition of these elements to the chromosome ends.

Recently much attention has been devoted to the protein moiety of the telomere, in studies performed mostly with *Saccharomyces cerevisiae* and in human cells. In fact, besides providing a terminal buffer zone that limits the effect of loss of DNA from the chromosome ends, telomeric DNA contains binding sites for proteins that bind DNA directly and which in turn can interact with other telomeric proteins. Some telomeric DNA binding proteins bind to the 3' single strand overhang on the chromosome ends, where they exert a protective capping function that prevents the cells from recognizing the ends as damaged DNA. The first of such capping proteins TEBP, was discovered in the ciliate *Oxytricha nova (13,14)*. In budding yeast a protein, Cdc13p, with no sequence similarity to TEBP, binds to the 3' DNA overhang and recruits two other capping proteins, Stn1p and Ten1p, to form a capping complex *(15,16)*. In fission yeast the Pot1 protein binds the 3' overhang and acts as a capping factor, and in mammals an ortholog of fission yeast Pot1 has a similar function *(17)*. Other telomeric proteins bind to the double-stranded DNA contained in the bulk of the telosome. The first such protein, ScRap1p, was discovered in *S. cerevisiae (18)*, and a similar protein (KlRap1p) was found in *Kluyveromyces lactis*

(19). In mammals two double-stranded DNA binding proteins are known, TRF1 and TRF2 *(20,21)*, and a protein with similar functions, Taz1, was found in fission yeast *(22)*. It has been shown that TRF2 causes a structural rearrangement of the end of the telomere, consisting in the formation of a large duplex loop in vivo (t loop), probably through strand invasion of the 3' overhang into the preceding telomeric tract *(23)*. T loops have been found at the termini of telomeres from a number of other organisms, and provide an additional protective and capping function. The study of telomeric proteins has shown that they recruit other proteins forming complexes that are involved in a number of functions carried out by the telomere, including and in addition to physical protection and replication.

First, the length of the telomere is regulated and is set at different species specific lengths. In budding yeast telomere length is controlled by the number of Rap1p molecules that bind at any given moment to the telomere *(24)*. It has been suggested that TRF1 and TRF2 act similarly in mammals *(25)*. Ever since it was discovered that human cells in culture can only divide a number of times ranging between 30 and 60, it was suggested that the shortening of telomeres in somatic tissues having inactive telomerase constituted a mitotic clock related to replicative senescence and in vivo human ageing. However, the connection between telomere loss and aging is not well understood; studies performed in mice in which telomerase activity had been knocked out has shown that complete lack of telomerase activity has damaging consequences, but only after more than six generations *(26)*.

Another important property of the telomere that has been studied extensively in budding yeast and in *Drosophila* is the telomeric position effect, which consists in the switching of subtelomerically located genes between silenced and nonsilenced states. Such states are semistable for several cell divisions and are produced by a transition between a euchromatic and a heterochromatic chromatin structure determined by the activity of telomere binding and telomere-associated proteins. In budding yeast the proteins involved are Rap1p and the Sir2-4 proteins, the yKu70-Ku80 heterodimer, and underacetylated histones in the subtelomeric nucleosomes *(27)*.

Telomeres also have a role in nuclear architecture. In yeast undergoing mitosis, telomeres are anchored by a protein complex to nuclear membrane pores *(28)*. Furthermore, there is evidence that the intranuclear spatial positioning of telomeres determines their transcriptional state *(29)*. At the beginning of meiosis in plants *(30)*, mammals *(31)*, and in fission yeast *(32,33)*, telomeres relocate to the nuclear envelope and cluster to a limited sector of the nuclear membrane. It is thought that this telomeric arrangement may facilitate pairing of chromosomes and homology searching, as well as synapsis *(30)*.

Telomere length is one of the most powerful parameters for the study of telomere function and, indirectly, of telomerase activity. Telomere length may be determined using molecular and cytological methods, the former making use of Southern hybridization to measure the most terminal restriction fragments, the latter being based on the identification of telomeres within the cells with the *in situ* hybridization procedure. In **Subheading 2.**, the two approaches will be described in detail.

Telomerase is a polymerase consisting of one RNA and several protein subunits. Only the RNA subunit (TER) and the catalytic subunit (TERT) are necessary for telomerase activity in vitro. The RNA contains a region complementary to the G-rich strand of the telomeric DNA; the TERT subunit reverse transcribes this sequence and adds the newly synthesized DNA repeats onto the 3' overhang of the telomere. A number of proteins are necessary for the in vivo activity of telomerase. In budding yeast the best studied case, at least three proteins have such an accessory role: est1p, est3p, and cdc13p and in their absence shortered telomeres are produced *(34,35)*.

It was postulated by Blackburn et al. that telomeres exchange their structure between an accessible and an inaccessible state to modulate telomerase activity. This switching could be stochastic or controlled by checkpoint signals such as kinases. Of particular interest is the finding that very short telomeres that usually cause growth arrest are still competent for telomerase activity in the absence of ataxia telangiectasia mutated (ATM) kinases, leading to a delay of replicative senescence *(36)*. This result suggests that ATM kinases may mediate the interaction between telomerase and its substrate, the telomere. Recently, a similar role was suggested for the human ATM kinase *(37)*.

Analysis of tissues and organs during development has revealed that telomerase is turned off from embryogenesis onwards, with few exceptions: basal/stem cell compartments of the immune system, skin, and intestine *(38)*. Switching off telomerase during development may determine the starting point of the telomere replicometer (telometer) a mechanism involved in determining the cellular life span in vitro and probably in vivo. Although in yeast it seems that the primary regulatory mechanism of telomerase activity is the telomere itself *(36)*, in multicellular organisms different mechanisms based on regulation of hTERT expression (silencing, alternative splicing, repression), stability of the holoenzyme and posttranslational modifications have been described. More recently the finding in human cells of a potent telomerase inhibitor (PinX1) opens the possibility that specific *trans*-acting factors may also contribute to regulate telomerase activity *(39)*.

Both basic and applied research on telomerase depends on the availability of reproducible and highly sensitive activity assays. One of the most popular methods to detect telomerase activity is the polymerase chain reaction (PCR)-based telomeric repeat amplification protocol (TRAP) assay, but other methods consisting in the direct analysis by gel electrophoresis of the DNA products extended by the telomerase have also been used. Because of the relationship between telomerase activity and proliferative capacity of the cells, telomerase is considered to be a tool to establish the malignant phenotype of tumor cells *(40,41)* and an oncology target *(42)*. In both cases the TRAP assay is used to evaluate tumor progression and to determine the efficacy of the therapeutic treatments. An alternative approach to TRAP is the quantitative analysis of hTERT mRNA, the catalytic subunit of telomerase *(43,44)*. This method is supported by the finding that hTERT expression correlates with telomerase activity *(45)*.

Telomerase assays consist in the identification of an RNA dependent synthesis of homogeneous telomeric DNA repeats that occurs by addition of a newly synthesized G-rich strand of the repeats to the 3' end of a telomeric DNA primer. The resulting products may be analyzed directly by polyacrylammide gel electrophoresis or after an amplification step performed by PCR (TRAP). The number of telomeric repeats added in vitro to the primer depends on the processivity of the enzyme: the yeast enzyme is non processive and a limited number of nucleotides, up to six, that corresponds to only one round of synthesis, is added to the primer *(46)*; protozoan and vertebrate telomerases are able to translocate many times to the beginning of their template region and place dozens of telomeric repeats on a single primer *(47)*. In this chapter, we describe in detail the TRAP assay *(48)* and the methods used to prepare the protein extracts from biological samples. For the conventional methods, which make use of fractionated extracts and do not use the PCR amplification step, a detailed description of the ciliate and yeast assays is reported in Greider and Blackburn *(49)*, Cohn and Blackburn *(50)*, and Counter et al. *(51)*.

2. Materials and Methods

2.1. Telomere Measurement

The following considerations apply to the the telomeric DNA belonging to the most common class of telomeres, comprised of a large number of short repeats.

2.1.1. Telomeric Restriction Fragment (TRF) Analysis

Telomere length can be measured by determining the size of the most terminal TRFs that span the distance between the telomere end and the most terminal restriction sites. TRFs are obtained by digesting genomic DNA with restriction enzymes that recognize 4-bp cutting sites. Because telomeres consist of repetitions of simple sequences, they lack the most common restriction sites, and therefore, following digestion, chromosomal DNA is cut into small fragments, whereas telomeric DNA remains uncut. An example is reported in **Fig. 1A.**

2.1.1.1. GENOMIC DNA ISOLATION

DNA can be isolated from tissue or cell cultures using standard protocols or commercial kits. It is recommended to analyze the quality of preparations of high molecular weight DNA by pulse field gel electrophoresis or conventional electrophoresis in 0.8% agarose gels. This parameter is particularly important if one is working with very long telomeres.

2.1.1.2. DIGESTION WITH 4-CUTTING ENZYMES

Approximately 10–15 (g DNA samples are digested with a mixture of two to four of the following enzymes: *Rsa*I, *Hinf*I, *Sau*3AI, and *Hae*III. The digestion is conducted in a 1X final concentration of enzyme buffer for 12–16 h at 37°C with 10–20 U of the enzyme pool. After digestion, DNA aliquots are quantified by standard methods (spectrophotometer, fluorimeter).

2.1.1.3. GEL ELECTROPHORESIS

Before use one aliquot of each sample is controlled by electrophoresis on a minigel: load about 1 µg of digested sample and separate DNA fragments at 5 V/cm for 1 h.

The cleaved DNA (at least 5 µg/sample) can be separated by conventional electrophoresis in 0.6%–0.8% agarose gels (5 V/cm for 5–6 h or 1 V/cm for 16–18 h for best resolution of long telomeres) or PFGE under conditions that separate 5–50 kb (the algorithm of the Bio–Rad CHEF-Mapper II can be used to set up the running conditions).

After electrophoresis, stain the gel, wash twice in distilled sterile water and photograph.

2.1.1.4. SOUTHERN TRANSFER AND HYBRIDIZATION

Southern transfer and hybridization are performed using conventional protocols as reported in Sambrook and Russell *(52)*. Hybridization can be performed using $(T_2AG_3)_{10}$ 5'-labeled oligo probes or cloned random-primed telomeric fragments (*see* **Note 1**).

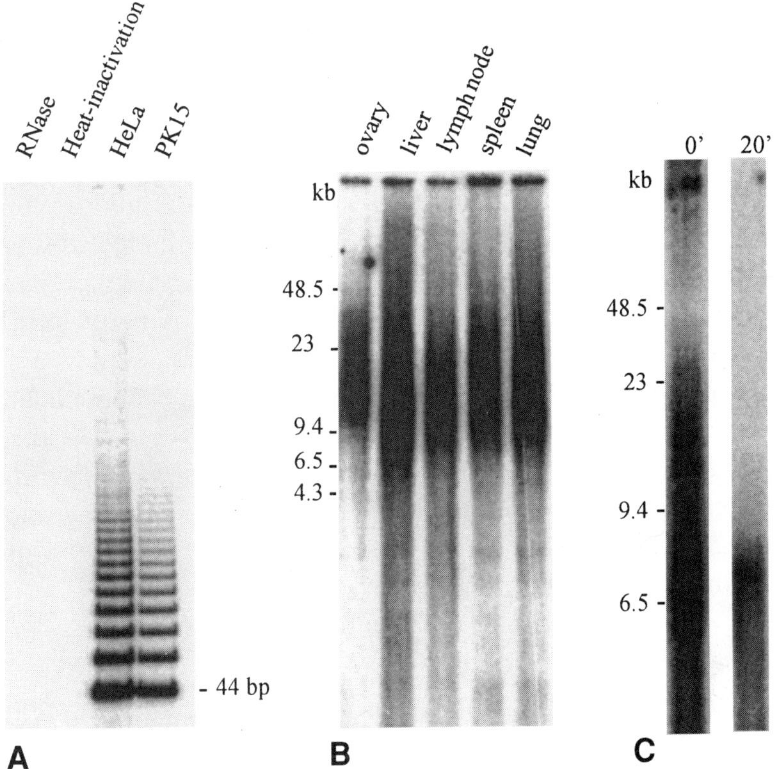

Fig. 1. (**A**) TRAP assay in human (HeLa) and pig (PK15) cells. Each lane contains the assay obtained with 5 µg NP40 extract. HeLa samples are assayed either untreated, pretreated with RNase or after heat inactivation. Products of telomerase start at 44 bp (TS/ACT primers) and display a 6-bp periodicity. (**B**) Telomere lengths in tissues of an adolescent female pig. Hybridization of *Hinf*I/*Sau*3AI-digested genomic DNA from the indicated tissues to the (TTAGGG)$_{10}$ probe. (**C**) *BAL*31 exonulease sensitivity of pig lymph node terminal restriction fragments. Genomic DNA was treated with Bal31 for 20 min and subsequently digested with *Hinf*I/*Sau*3AI. The autoradiogram shows that the bulk of the T$_2$AG$_3$ hybridization signal (6–20 kb) was significantly reduced in size at the 20-min time point with respect to the *BAL*31-undigested sample (0 min). Undigested λ DNA and λ-*Hin*dIII-size markers are reported on the left. All three panels contain unpublished results by the authors.

In many organisms, telomeric repeats are not only located at the chromosomal ends but also at a number of intrachromosomal sites *(53)*. A very useful tool for distinguishing between the two types of repeats, as well as for determining the position of specific sequences in the subtelomeric region of chromosomes is the enzyme BAL31. The activity of this nuclease on DNA is predominantly exonucleolytic, causing both strands of double-stranded DNA to be shortened from both ends. BAL31 treatment of restricted genomic DNA progressively removes the TRFs from the southern blots (**Fig. 1A**), and the bands that remain in the lanes correspond to intrachromosomal sites. In some cases, such as in Chinese hamster, the intrachromosomal repeats are so abundant as to render the terminal truly telomeric DNA invisible in Southern analysis *(54)*. In these cases, truly telomeric sequences can be visualized by fluorescence *in situ* hybridization and quantified by q-FISH and Fluo-FISH (*see* **Subheading 2.1.2.**).

2.1.2. Q-FISH and Fluo-FISH

Telomeres can be detected directly inside the cells using fluorescent-labeled probes and the *in situ* hybridization technique. This approach is common to q-FISH *(55)* and Fluo-FISH *(56,57)*, which are both based on the notion that after quantitative fluorescent *in situ* hybridization, the intensity of the fluorescent signals is proportional to the amount of telomeric DNA to which the probe is bound, thereby furnishing an indirect measurement of telomere length. Q-FISH consists of a canonical FISH performed with a Cy3-conjugated peptide nucleic acid probe. After hybridization, the chromosomes are counterstained with DAPI. Cy3 and DAPI images are captured separately using a dedicated computer program (TFL-TELO). Using the same program, the images are processed and integrated to furnish the fluorescence intensity value of individual telomeres. It was estimated that an average of 30 metaphases are sufficient to obtain a good indication of telomere length (SD of 10–15%). The complete procedure, software included, is available in a commercial kit distributed by DAKO. Fluo-FISH overcomes some of the limitations of q-FISH, especially those related to the processing time and the necessity to analyze multiple metaphases for each sample. Fluo-FISH can be completed within 1 d and allows the acquisition of data on telomere length from single cells and from subsets of cells (i.e., lymphocytes and granulocytes) present in the sample under investigation. Similarly to q-FISH, the procedure is based on quantitative hybridization of a fluorescent-conjugated telomere-peptide nucleic acid probe to the sample, but the acquisition and analysis of telomeric signals are performed with a flow cytometer. The protocol can be subdived into six steps: (1) cell separation/preparation; (2) DNA denaturation; (3) hybridization; (4) washes; (5) DNA counterstaining; and (6) data acquisition and flow cytometer analysis. Each of these steps needs to be standardized to obtain reliable and accurate results, and for this reason the authors have reported the relevant parameters affecting the procedures and the pitfalls that can arise *(58)*.

Q-FISH has a very high resolution, up to 200 bp, and can be performed with genomes containing interstitial telomeric sites, such as Chinese hamster. Fluo-FISH offers the advantages of a faster procedure and the possibility of obtaining data at both single-cell and subpopulation levels.

2.2. Telomere Cloning

Eukaryotic organisms can recognize and use telomeres from distantly related organisms *(60,61)*. Telomeres transferred between organisms having different telomeric sequences produce hybrid telomeres, containing medially located telomeric sequences characteristic of the donor, to which the recipient organism adds its typical telomeric sequence. This has made possible the functional cloning of yeast telomeres in yeast itself *(59)* and of mammalian telomeres and subtelomeric DNA *(62,63)*. Today telomeres are cloned directly by molecular biological techniques as outlined below.

2.2.1. PTA-Cloning of Telomeres

To isolate telomeres and their adjacent sequences (telomere-associated sequences: TASs) Wang and collaborators developed a method called primer-tagged amplification *(64)*. Partially degenerated random primers (6-MW, *see* **Table 1**) are used to generate primer-tagged genomic DNA fragments that are subsequently amplified with PCR. In the PCR reaction, the primed-tagged genomic DNA fragments that also contain primed tagged-telomeric DNA are amplified using 6-MW telomeric primers (telo32 and telo18, *see* **Table 1**). All the primers contain the *Xho*I site subsequently used for cloning. The experimental procedure is performed in four steps.

First step: random-primer annealing to genomic DNA. 1 µg 6-MW and 10–50 µg genomic DNA are mixed in a total volume of 80 µL water. The mixture is denatured in boiling water for 5 min and quickly cooled (room temperature for 2 min, on ice for at least 2 min).

In the second step, synthesis of random-primed tagged DNA fragments is performed. For a volume reaction of 100 µL: 1X Klenow buffer, 0.2 m*M* dNTP, 3 U of Klenow enzyme, 80 µL annealing reaction (from **step 1**). Incubate the mixture at 37°C for 30 min. To prolong the strand extension, add 0.5 U of Taq polymerase and incubate at 72°C for 15 min. Proceed to phenol/chloroform 1:1 extraction and ethanol precipitation adding only 2.5 vol of 100% ethanol and leave at room temperature for 5 min. Collect DNA by standard centrifugation, wash with 1 vol of 70% ethanol and air dry. Resuspend DNA in 50 µL water.

The third step is PCR. For cloning T_2AG_3 telomeres we have successfully used telo32 and telo18 in combination with 6-MW. Telo32 and telo18 are used as a mixture with a molar ratio of 10:1. This combination avoids slippage of primer on tandem telomeric repeats. The PCR reaction is performed in 50 µL containing 1X PCR buffer, 0.2 m*M* dNTP, 10–20 ng of DNA template, 100 ng of 6-MW, 10 ng telo32, 100 ng telo18, and 1.5 U Taq polymerase. The PCR conditions are as follows: 94°C for 30 s, 54°C for 1 min, 72°C for 1.5 min. Control the PCR products with agarose gels and eventually purify the fragments of the expected size from the gels.

In the fourth step, cloning, digest PCR products with *Xho*I and ligate to a *Xho*I-linearized vector. It is important to choose a low copy number vector and to use recombination-deficient bacterial strains to minimize rearrangements of the telomeric inserts. In our

Table 1
TRAP and PCR Primers

Name	Sequence	Ta[a]
TS	AATCCGTCGAGCAGAGTT	
CX	CCCTTACCCTTACCCTTACCCTAA	50
Cxa	GTGTAACCCTAACCCTAACCC	50
ACX	GCGCGGCTTACCCTTACCCTTACCCTAACC	60
ACT	GCGCGGCTAACCCTAACCCTAACC	60
TSNT	AATCCGTCGAGCAGAGTTAAAAGGCCGAGAAGCGAT	60[b]
NT	ATCGCTTCTCGGCCTTTT	60[b]
6-MW	CCGACTCGAGNNNNNNTGTGG	54
telo32	GACTGACTCGAGCTAACCCTAACCCTTAACCCT	54
telo18	GACTGACTCGAGCTAACC	54

[a]Annealing temperature of the indicated primer in combination with TS.
[b]To be used in the presence of TS/ACX or TS/ACT.

hands, good results were obtained with plasmid pACYC177 and DH10B cells.

2.2.2. PCR Cloning of Telomeres

A similar approach, based on PCR amplification of specific telomeric DNA was developed to clone yeast telomeres *(65)*. The procedure is based on addition of a primer (anchor) to the 3' telomeric overhang, and subsequent amplification of telomeres with a primer complementary to the anchor (hybridization anchor) and a primer complementary to the subtelomeric sequences. This procedure with respect to the previous one requires that the sequences immediately upstream to the telomeric repeats be known, and it is therefore essentially used to control the sequence and to determine the exact length of previously mapped telomeres.

2.3. TRAP Telomerase Assay

The TRAP assay *(48)* can be conducted using protein extracts prepared from a variety of biological samples by detergent lysis in buffer containing CHAPS or NP40. The assay consists of two steps: (1) the telomerase reaction that extends the DNA primer and (2) the PCR that amplifies the extended products. In the first step, telomerase binds and elongates the TS **(Table 1)** primer by adding multiple telomeric repeats that produce a 6 nt ladder. In the second step the products are amplified by PCR using TS and a reverse primer (CX in the original assay, **Table 1**). An example is reported in **Fig. 1B**. The specificity of the reaction is determined by its sensitivity to RNAse and proteinase treatments. The nucleotide requirement of the first step is determined by the sequence of the telomeric repeat (for T_2AG_3 telomeres dCTP is not required).

2.3.1. Preparation of Protein Extracts and Protein Quantification

From cultured cells, wash and scrape off cells carefully with sterile phosphate-buffered saline (PBS). Harvest cells into a 15-mL sterile tube and centrifuge at 200*g* for 5 min. Wash twice with PBS (*52*). The pellet can be stored at –80°C (remove all PBS before freezing) until use (up to one year) or proceed immediately to the next step. Resuspend cells in NP40 or CHAPS lysis buffer (**Table 2**). Add 200 mL of 1X lysis buffer per 10^7 cells; if cells are not counted, use 50 mL for a small tissue culture plate (60 mm) and 100 mL for a medium plate (10 cm). Once the pellet is resuspended, incubate on ice for 30 min, then spin at 12,000*g* for 30 min. Collect the supernatant without disturbing the pellet, dispense 20-mL aliquots into 0.5 mL RNAse-free tubes. Use one sample to determine the protein concentration (*see* **Subheading 2.3.2.**). Quick freeze the remaining aliquots in liquid N_2 and store at –80°C for up to 6 mo. After use, aliquots should be rapidly frozen again, but should not be used more than five times. Do not exceed in the number of extracts to prepare (10–12 at a time) in order to avoid a long period of standing on ice.

Tissue can be rapidly homogenized by a mechanical homogenizer in lysis buffer (CHAPS or NP40) or powdered in liquid N_2 using a mortar and pestle. After tissues have been powdered, transfer them to a preweighed tube and add lysis buffer. If multiple samples are processed, powder all of them first and keep in liquid N_2, then add lysis buffer to all of them. The lysis buffer is added at 200 mL/100 mg of tissue. Proceed as described for cultured cells.

2.3.2. Determination of the Protein Concentration in the Extracts

Protein concentration is measured against bovine serum albumin (BSA) using a Bio-Rad protein assay reagent (Bio-Rad, cat. no. 500-0006). Prepare various concentrations of BSA standard in lysis buffer taking into account that the standard curve remains linear from only approx 2.5–15 mg of BSA. If the absorbance values of the protein samples fall outside of this range, the margin of error becomes very high. Prepare various dilutions of the extracts (5–100 mL) in a total volume of 100 mL lysis buffer, add 1 mL protein reagent and vortex. After 2 min incubation, but before 1 h, determine the OD_{595}.

**Table 2
Buffers**

Final volume 100 mL

Stock	Final concentration	Final volume
CHAPS lysis buffer		
1 *M* Tris-HCl pH 7.5	10 m*M*	1 mL
1 *M* MgCl2	1 m*M*	100 μL
100 m*M* EGTA	1 m*M*	1 mL
100% glycerol	10%	10 mL
RNAse-free water		87.9 mL

Autoclave, store at 4°C. Just prior to use, add to 950 μL of Lysis buffer:

0.35 μL 14,4*M* β-mercaptoethanol (BME)
50 μL 10% CHAPS

Stock	Final concentration	Final volume
NP40 lysis buffer		
(Final volume 100 mL)		
Final volume, 100 mL		
1 *M* Tris-HCl pH 7.5	10 m*M*	1 mL
1 *M* MgCl$_2$	1 m*M*	100 μL
100 m*M* EGTA	1 m*M*	1 mL
100% glycerol	10%	10 mL
2.5 *M* NaCl	150 m*M*	6 *mL*
RNAse-free water		80.74 *mL*

Autoclave, store at 4°C. Just prior to use, add to 988.4 μL of lysis buffer:

1 μL 0.1 *M* PM*SF*
0.35 μL 14.4 *M* β-mercaptoethanol (BME)
0.2 μL 0.12 *M* deoxycholate sodium salt
10 μL 100% NP40

Stock	Final concentration	Final volume
10X TRAP buffer		
(Final volume 5 mL)		
1 *M* Tris-HCl, pH 8.3	200 m*M*	1 mL
1 *M* MgCl$_2$	1 m*M*	100 μL
1 *M* KCl	680 m*M*	3.4 mL
100% Tween 20	0.5%	25 μL
100 m*M* EGTA	10 m*M*	500 μL

Mix and filter trough 0.45-μm pore filters. Store aliquots at −20°C for up to 6 mo

2.3.3. TRAP Reaction

The first step of the assay enables the telomerase present in the extracts to add telomeric repeats onto the 3' end of the TS primer; subsequently the products are amplified by PCR in the presence of TS and reverse primer (**Table 1**). Telomerase activity results in a 6-bp ladder when the PCR products are separated on a 10–12% polyacrylamide nondenaturing gel.

The reactions are set up in PCR tubes in a final volume of 50 mL. Each reaction contains 1X TRAP buffer (**Table 2**); 2.5 mM each dNTP; 150 ng of TS; 2.5 U Taq polymerase; 300 ng reverse primer Cxa, ACX or ACT); the internal control (IC) TSNT/NT (0.7 fg/0.3 mg); and 5–10 mg protein extract (**Table 1, Notes 2–4**).

The telomerase products can be detected using either a nonradioactive or radioactive TS primer. In the radioactive method the TS primer is end-labeled with ^{32}P as follows: 1 mg TS; 25 mCi [^{32}P]γ-ATP; 1X T4 kinase buffer; 10 U T4 kinase enzyme; reaction volume, 10 mL. Incubate the reaction at 37°C for 30 min. Stop the reaction at 65°C for 15 min and store at –20°C until needed. For radioactive detection use 20 ng hot TS and 130 ng cold TS for each reaction; if non radioactive detection is used add 150 ng TS.

PCR conditions are 27 cycles at 94°C for 1 min (denaturation); annealing temperature (Ta) for 30 s (annealing) and 72°C for 30 s (elongation). Ta may be different according to the primers used in the reaction (*see* **Table 1**).

2.3.4. Gel Electrophoresis and Detection

Products can be resolved by electrophoresis in a 10–15 cm vertical nondenaturating 12% polyacrylamide gel (19:1 w/w acrylamide/ N,N'-methylene bisacrylamide) in 0.5X TBE. Add 10 mL of 6X DNA loading dye and load the samples onto the gel. Run at 400–500 V until the second dye is approx 3 cm from the bottom. The gels are dried for 1 h at 80°C. For the nonradioactive assay stain the gel in a 1/10,000 dilution of SYBR Green in TBE 0.5X for 30 min. To visualize the telomeric products use a UV source with a SYBR green photographic filter. Alternately, use ethidium bromide as a DNA dye, although it will not be as sensitive as SYBR green. A third nonradioactive method was developed by Wen and collaborators *(66)* which makes use of silver staining of the acrylamide gels containing the TRAP products. Briefly fix the gel in 0.5% acetic acid, 10% ethanol for 15 min, add 0.2% $AgNO_2$ for 10 min, wash twice in ddH_2O, incubate in 0.1% formaldehyde, 3% NaOH for 10 min, photograph (keep the gel wet) on 677 polaroid film *(67)*. For the radioactive assay the gel is analyzed in the PhosphorImager or other similar device, after a 5- to 24-h exposure of a PhosphoImager screen, or exposure to X-ray films at –80°C for a time period that depends on the strength of the radioactive signal. The IC appears as a 36-bp band, and the size of the smallest telomerase product is determined by the primers used: 40 bp with TS/CX, 44 bp with TS/ACT and 50 bp with TS/ACX.

3. Notes

1. After electrophoresis, the telomeric restriction fragments are well separated from the bulk of the chromosomal DNA which usually runs as 500- to 2000-bp fragments. This phenomenon, which indicates complete digestion, is clearly visible after hybridization, because telomeric signals localize to the portion of the gel that is apparently free of DNA (this is due to the very low amount of telomeric DNA with respect to the total genome).

2. It is important to keep everything cold (0°C), to use RNase free materials, and to observe the general operating rules for handling RNA containing materials. To minimize cross-contamination, the working area for extract preparations should be separated from the area used to set up the TRAP reactions. Moreover, it is recommended to store all solutions separately from other laboratory reagents.

3. The original assay makes use of the substrate primer TS and the reverse primer CX. CX can cause a number of artifacts because of primer dimer products with TS and binding with telomerase products *(48)*. The latter will cause elongation of telomerase products (staggered annealing) and affect the processivity (the number of telomeric repeats added to the telomeric primer) of the reaction. These undesired reactions may be partially eliminated by using different reverse primers: CXa, not complementary to the TS 3' end and containing a three nucleotide 5' anchor not complementary to the 3'

end of the telomerase products, that should reduce staggered annealing *(66)*; ACX, similarly to CXa, contains a 6 nucleotide 5' anchor and a permutation of the telomeric repeat that significantly reduces artifact products *(67)*. All the primers are reported in **Table 1**.

4. The TRAP IC was developed to control the PCR step of the assay *(69)*, it consists of the TSNT standard amplified by TS and the dedicated primer NT, which is not a substrate for telomerase (**Table 1**). IC produces a 36-bp band coamplified proportionally with telomerase products (if inhibitors are present in the extracts, both telomerase and IC products are affected). IC is particularly useful for measuring telomerase activity in clinical samples since it allows accurate quantification of the enzyme activity and comparison between the activity present in different extracts.

5. TRAP controls. RNase treatment: telomerase is sensitive to RNase because of the RNA component of the enzyme which provides the template for the synthesis of telomeric repeats; 10–30 mg extracts are incubated with 30 mg RNaseA for 25 min at 30°C. Heat inactivation: heat treatment destroys telomerase activity; the assay is conducted by incubating a small aliquot (10–30 mg) of the extracts at 85°C for 10 min. The treated samples are subsequently used in a standard TRAP assay; only the IC band should be visible in these reactions. Positive control: it is recommended to use a cellular extract from telomerase positive cell lines (HeLa, 293) in order to control that all the operating conditions are optimal for detection of telomerase activity in the experimental samples.

Acknowledgment

The unpublished results presented in **Fig. 1** are from work supported by a grant (Modalità alternative di transgenesi animale; ricerche scientifiche sulla costruzione di mini e microcromosomi artificiali) from the Italian Ministry of University and Research.

References

1. Ishikawa, F. and Naito, T. (1999) Why do we have linear chromosomes? A matter of Adam and Eve. *Mutat. Res.* **434,** 99–107.
2. Muller, H. J. (1938) The remaking of chromosomes. *Collect. Net.* **8,** 182–195.

3. Muller, H. J. (1940) An analysis of the process of structural change in the chromosomes of Drosophila. *J. Genet.* **40,** 1–66.

4. McClintock, B. (1939) The behavior in successive nuclear divisions of a chromosome broken at meiosis. *Proc. Natl. Acad. Sci. USA* **25,** 405–416.

5. Wright, J. H., Gottschling, D. E., and Zakian, V. A. (1992) *Saccharomyces* telomeres assume a nonnucleosomal chromatin structure. *Genes. Dev.* **6,** 197–210.

6. Blackburn, E. H. and Gall, J. G. (1978) A tandemly repeated sequence at the termini of the extrachromosomal ribosomal RNA genes in *Tetrahymena. J. Mol. Biol.* **120,** 33–53.

7. Klobutcher, L. A., Swanton, M. T., Donini, P., and Prescott, D. M. (1981) All gene–sized DNA molecules in four species of hypotrichs have the same terminal sequence and an unusual 3' terminus. *Proc. Natl. Acad. Sci. USA* **78,** 3015–3019.

8. Lundblad, V. and Blackburn, H. (1993) An alternative pathway for yeast telomere maintenance rescues est1-senescence. *Cell* **73,** 347–360.

9. McEachern, M. J. and Blackburn, H. (1996) Cap-prevented recombination between terminal telomeric repeat arrays (telomere CPR) maintains telomeres in Kluyveromyces lactis lacking telomerase. *Genes Dev.* **10,** 1822–1834.

10. Yeager, T. R., Neumann, A. A., Englezou, A., Huschtscha, L. I., Noble, J. R. et al. (1999) Telomerase-negative immortalized human cells contain a novel type of promyelocytic leukemia (PML) body. *Cancer Res.* **59,** 4175–4179.

11. Saiga, H. and Edstrom, J. E. (1985) Long tandem arrays of complex repeat units in Chironomus telomeres. *EMBO J.* **79,** 315–328.

12. Roth, C. W., Kobeski, F., Walter, M. F., and Biessmann, H. (1997) Chromosome end elongation by recombination in the mosquito *Anopheles gambiae. Mol. Cell. Biol.* **17,** 5176–5183.

13. Gottschling, D. E. and Cech, T. R. (1984) Chromatin structure of the molecular ends of Oxytricha macronuclear DNA: phased nucleosomes and a telomeric complex. *Cell* **38,** 501–510.

14. Gottschling, D. E. and Zakian, V. A. (1986) Telomere proteins: Specific recognition and protection of the natural termini of Oxytricha macronuclear DNA. *Cell* **47,** 195–205.

15. Pennock, E., Buckley, K. and Lundblad, V. (2001) Cdc13 delivers separate complexes to the telomere for end protection and replication. *Cell* **104,** 387–396.

16. Grandin, N., Damon, C. and Charbonneau, M. (2001) Ten1 functions in telomere end protection and length regulation in association with Stn1 and Cdc13. *EMBO J.* **20,** 1173–1183.

17. Baumann, P. and Cech, T. R. (2001) Pot1, the putative telomere end-binding protein in fission yeast and humans. *Science* **292,** 1171–1175.

18. Berman, J. and Tachibana, C. Y. (1986) Identification of a telomere-binding activity from yeast. *Proc. Natl. Acad. Sci. USA* **83,** 3713–3717.

19. Larson, G. P., Castanotto, D., Rossi, J. J. and Malafa, M. P. (1994) Isolation and functional analysis of a *Kluyveromyces lactis* RAP1 homologue. *Gene* **150,** 35–41.

20. Bilaud, T., Brun, C., Ancelin, K., Koering, C. E., Laroche, T. and Gilson, E. (1997) Telomeric localization of TRF2, a a novel human telobox protein. *Nat. Genet.* **17,** 236–239.

21. Broccoli, D., Smogorzewska, A., Chong, L., and De Lange, T. (1997) Human telomeres contain two distinct Myb-related proteins, TRF1 and TRF2. *Nat. Genet.* **17,** 231–235.

22. Ferreira, M. and Cooper, J. (2001) The fission yeast Taz1 protein protects chromosomes from Ku-dependent end-to-end fusion. *Mol. Cell* **7,** 55–63.

23. Griffith, J. D., Comeau L., Rosenfield, S., Stansel R. M., Bianchi A., Moss, H., et al. (1999) Mammalian telomeres end in a large duplex loop. *Cell* **97,** 503–514.

24. Marcand, S., Wotton, D., Gilson, E. and Shore, D. (1997) Rap1p and telomere length regulation in yeast. *Ciba Found. Symp.* **211,** 76–93.

25. Smogorzewska, A., van Steensel, B., Bianchi, A., Oelmann, S., Schaefer, M. R., et al. (2000) Control of human telomere length by TRF1 and TRF2. *Mol. Cell. Biol.* **20,** 1659–1668.

26. Lee, H. W., Blasco, M. A., Gottlieb, G. J., Horner, J. W. II, Greider, C. W. and Harley, C. B. (1998) Essential role of mouse telomerase in highly proliferative organs. *Nature* **362,** 569–574.

27. Lustig, A. J. (1998) Mechanisms of silencing in *Saccharomyces cerevisiae*. *Curr. Opin. Genet. Dev.* **8,** 233–239.

28. Galy, V., Olivo-Merin, J. C., Scherthan, H., Doye V., Rascalou, N. and Nehrbass, U. (2000) Nuclear pore complexes in the organization of silent telomeric chromatin. *Nature* **403,** 108–112.

29. Feuerbach, F., Galy, V., Sticken, E. T., Fromont-Racine, M., Jacquier, A., Gilson, E., et al. (2002) Nuclear architecture and spatial positioning help establish transcriptional states of telomeres in yeast. *Nat. Cell Biol.* **4,** 214–221.

30. Bass, H. W., Riera-Lizarazu, O., Ananiev, E. V., Bordoli, S. J., Rines, H. W., Phillips, R. L., et al. (2000) Evidence for the coincident initiation of homolog pairing and synapsis during the telomere clustering (bouquet) stage of meiotic prophase. *J. Cell Sci.* **113,** 1033–1042.

31. Loidl, J. (1990) The initiation of meiotic chromosome pairing: the cytological view. *Genome* **33,** 759–778.

32. Cooper, J. P. (2000) Telomere transitions in yeast: the end of the chromosome as we know it. *Curr. Opin. Genet Dev.* **10,** 169–177.

33. Nimmo, E. R., Pidoux, A. L., Perry, P. E., and Allshire, R. C. (1998) Defective meiosis in telomere-silencing mutants of Schizosaccharomyces pombe. *Nature* **392,** 825–828.

34. Lingner, J., Cech, T. R., Hughes, T. R. and Lundblad, V. (1997) Three ever shorter telomere (EST) genes are dispensable for in vitro yeast tclomerase activity. *Proc. Natl. Acad. Sci. USA* **94,** 11,190-11,195.

35. Nugent, C. I., Hughes, T. R., Lue, N. F., and Lundblad, V. (1996) Cdc13p: a single-strand telomeric DNA-binding protein with dual role in yeast telomere maintenance. *Science* **274,** 249–252.

36. Chan, S. W., Chang, J., Prescott, J., and Blackburn, E. H. (2001) Altering telomere structure allows telomerase to act in yeast lacking ATM kinases. *Curr Biol.* **11,** 1240–1250.

37. Kishi, K. and Lu, P. L. (2002) A critical role for Pin2/TRF1 in ATM-dependent regulation. *J. Biol. Chem.* **277,** 7420–7429.

38. Forsyth, N. R., Wright, W. E and Shay, J. W. (2002) Telomerase and differentiation in multicelluar organisms: turn it off, turn it on, and turn it off again. *Differentiation* **69,** 188–197.

39. Zhou, X. Z. and Lu, K. P. (2001) The Pin2/TRF1-interacting protein PinX1 is a potent telomerase inhibitor. *Cell* **107,** 347–359.

40. Shen, Z.Y., Xu, L.Y., Li, C., Cai, W. J., Shen, J., Chen, J. Y., and Zeng, Y. (2001) A comparative study of telomerase activity and malignant phenotype in multistage carcinogenesis of esophageal epithelial cells induced by human papillomavirus. *Int. J. Mol. Med.* **8,** 633–639.

41. Muller, M. (2002) Telomerase: its clinical relevance in the diagnosis in bladder cancer. *Oncogene* **21,** 650–655.

42. Corey, D. R. (2002) Telomerase inhibition, oligonucleotides, and clinical trials. *Oncogene* **21,** 631–637.

43. Ito, H., Kio, S., Kanaya, T., Takakura, M., Inoue, M., and Namiki, M. (1998) Expression of human telomerase subunits and correlation with telomerase activity in urothelial cancer. *Clin. Cancer. Res.* **4,** 1603–1608.

44. de Kok, J. B., Schalken, J. A., Aalders, T. W., Ruers, T.J., Willelms, H. L., and Swinkels, D. W. (2000) Quantitative measurement of telomerase reverse transcriptase (hTERT) mRNA in urothelial cell carcinomas. *Int. J. Cancer* **87,** 217–220.

45. Nakamura, T. M, Morin, G. B., Chapman, K. B., Weinrich, S. L., Andrews, W. H., Lingner, J., et al. (1997) Telomerase catalytic subunit homologs from fission yeast and human. *Science* **277,** 955–959.

46. Prescot, J. and Blackburn, E. H. (1997) Functionally interacting telomerase RNAs in the yeast telomerase complex. *Genes Dev.***11,** 2790–800.

47. Lee, M. S. and Blackburn, E. H. (1993) Sequence-specific DNA primer effects on telomerase polymarization activity. *Mol. Cell. Biol.* **13,** 6586–6599.

48. Kim, N. W., Piatyszek, M. A., Prowse, K. R., Harley, C.B., West, M. D., Ho, P. L., et al. (1994) Specific association of human telomerase activity with immortal cells and cancer. *Science* **266,** 2011–2015.

49. Greider, C. W. and Blackburn, E., H. (1987) The telomere terminal transferase of Tetrahymena is a ribonucleoprotein enzyme with two kinds of primer specificity. *Cell.* **51,** 887–898.

50. Cohn, M. and Blackburn, H. E. (1995) Telomerase in yeast. *Science* **269,** 396–400.

51. Counter, C. M., Meyerson, M., Eaton, E. N. and Weinberg, R. A. (1997) The catalytic subunit of yeast telomerase. *Proc. Natl. Acad. Sci. USA* **94,** 9202–9207.

52. Sambrook, J. and Russell, D. W. (2001) *Molecular Cloning, a laboratory manual*, 3rd ed., Cold Spring Harbor Laboratory Press, Cold Spring Harbor, New York.

53. Meyne, J., Baker, R. J., Hobart, H. H., Hsu, T. C., Ryder, O. A., Ward, O. G., et al. (1990) Distribution of non-telomeric sites of the (TTAGGG)n in telomeric sequence in vertebrate chromosomes. *Chromosoma* **99,** 3–10.

54. Faravelli, M., Azzalin, C. M., Bertoni, L., Chernova O., Mondello, C., and Giulotto, E. (2002) Molecular organization of internal telomeric sequences in Chinese hamster chromosomes. *Gene* **283,** 11–16.

55. Poon, S. S. S., Martens, U. M., Ward, R. K., Lansdorp, P. M. (1999) Telomere length measurements using digital fluorescence microscopy. *Cytometry* **36,** 267–278.

56. Rufer, N., Dragowska, W., Thornbury, G., Roosnek, E., Lansdorp, P. M. (1998) Telomere length dynamics in a human lymphocytes subpopulations measured by flow cytometry. *Nature Biotech.* **16,** 743–747.

57. Hultdin, M., Gronlund, E., Norrback, K.-F., Ericsson-Lindstroem, E., Just, T. and Roos, G. (1998) telomere analysis by fluorescence in situ hybridization and flow cytometry. *Nucleic Acids Res.* **26,** 3651–3656.

58. Baerlocher, G. M., Mak, J., Tien, T., and Lansdorp, P. M. (2002) Telomere length measurements by fluorescence in situ hybridization and flow cytometry: Tips and pitfalls. *Cytometry* **47,** 89–99.

59. Szostak, J. W. and Blackburn, E. H. (1982) Cloning yeast telomeres on linear plasmid vectors. *Cell* **29,** 245–255.

60. Guerrini, A. M., Ascenzioni F., Tribioli C., and Donini P. (1985) Transformation of *Saccharomyces cerevisiae* and *Schizosaccharomyces pombe* with linear plasmids containing 2 micron sequences. *EMBO J.* **4,** 1569–1573.

61. Ascenzioni, F. and Lipps, H. J. (1986) A linear shuttle vector for yeast and hypotrichous ciliate Stylonychia. *Gene* **46,** 123–126.

62. Brown, W. R. (1989) Molecular cloning of human telomeres in yeast. *Nature* **338,** 774–776.

63. Guerrini, A. M., Ascenzioni, F., Pisani, G., Rappazzo G., Della Valle, G., and Donini, P. (1990) Cloning a fragment from the telomere of the long arm of human chromosome 9 in a YAC vector. *Chromosoma* **99,** 138–142.

64. Fu, G. and Barker, D. C. (1998) Rapid cloning of telomere-associated sequence using primer-tagged amplification. *Biotechniques* **24,** 386–389.

65. Tzfati, Y., Fulton, T. B., Roy, J. and Blackburn, E. H. (2000) Template boundary in a yeast telomerase specified by RNA structure. *Science* **288,** 863–867.

66. Wen, J. M., Sun, L. B., Zhang, M., and Zheng, M. H. (1998) A nonisotopic method for the detection of telomerase activity in tumor tissue: TRAP-silver staining assay. *J. Clinic. Pathol.* **51,** 110–112.

67. Dalla Torre, C. A., Maciel, R. M. B., Pinheiro, N. A., Andrade, J. A. D., de Toledo, S. R. C., et al.. (2002) TRAP-silver staining, a highly sensitive assay for measuring telomerase activity in tumor tissue and cell lines. *Bras. J. Med. Biol. Res.* **35,** 65–68.

68. Falchetti, M. L., Levi, A., Molinari, P., Verna, R., and D'Ambrosio, E. (1998) Increased sensitivity and reproducibility of TRAP assay by avoiding direct primers interaction. *Nucleic Acids Res.* **26,** 862–863.

69. Kim, N. W. and Wu, F. (1997) Advances in quantification and characterization of telomerase activity by the telomeric repeat amplification protocol (TRAP). *Nucleic Acids Res.* **25,** 2595–2597.

8

Telomerization of Mammalian Cells and Transplantation of Telomerized Cells in Immunodeficient Mice

Peter J. Hornsby

1. Introduction

Telomerization is an operational term for appparent immortalization sustained by prevention of telomere shortening. It is always necessary to use the term apparent because immortalization, if taken literally, requires one to demonstrate that cells divide forever, a property that is obviously impossible to verify. In practice, for human cells, division for over 200 times is generally accepted as indicating that cells are immortalized. It has been known since the pioneering experiments of Leonard Hayflick in the 1960s that the limited replicative capacity of human cells in culture is very unlikely to be an experimental artifact but is a reproducible biological phenomenon *(1)*. However, not until the discovery that the limitation in replicative capacity directly correlates with telomere shortening was the notion that it might be a culture artifact finally laid to rest *(2,3)*.

From: *Methods in Molecular Biology, Vol. 240:*
Mammalian Artificial Chromosomes: Methods and Protocols
Edited by: V. Sgaramella and S. Eridani © Humana Press Inc., Totowa, NJ

Telomeres shorten in most dividing human somatic cells because of the lack of telomerase activity that is required for telomere mainte-nance *(4,5)*. The lack of telomerase activity results from the absence of expression of the reverse transcriptase subunit (TERT) of the telomerase ribonucleoprotein complex *(6,7)*. When cells divide in the absence of telomerase activity, about 40–100 bp of the terminal telomeric repeat DNA is not replicated *(4,5)*. This amount is a con-stant for various types of human cells, thus providing a kind of mitotic counter *(4,5)*. Telomere shortening caused by lack of telomerase activity also occurs in many cell types in human tissues in an age-related fashion *(8)*.

After a normal human cell has divided a certain number of times, that number varying with the specific cell type and culture condi-tions, the telomeres become so short that they trigger a cell cycle checkpoint that puts the cell into a terminally nondividing state. This state has commonly been termed cellular senescence or replicative senescence. The block to proliferation in replicative senescence is similar to that caused by double-strand breaks in cellular DNA *(5)*, although whether this is actually the mechanism by which short telomeres are recognized is not clear. Further cell division is then blocked by inhibitors of cell proliferation, such as p21[SD11/WAF1/CIP1] and p16[INK4A] *(9,10)*.

Proof that the limitation on indefinite cell division in most human cells results from lack of expression of TERT was obtained by show-ing that forced expression of TERT is sufficient to immortalize nor-mal human fibroblasts and retinal pigmented epithelial cells. Human fibroblasts were the first cell type to be immortalized with telomerase and these cells have to date achieved a population dou-bling level (PDL) of 500 (they were at PDL 70 when hTERT was introduced and they do not normally divide beyond PDL 90) *(11)*. Immortalization was accompanied by increased or stabilized telom-ere length, but cells retain a normal karyotype *(12,13)*.

Beginning with the first reports of telomerization, it was specu-lated that this technology could be used to expand populations of cells for subsequent therapeutic transplantation *(11,13–15)*. This was thought of as particularly important for the replacement of tissues and organs damaged during aging *(16,17)*. In one proposed form of this therapy, cells with shortened telomeres would be isolated from a patient and telomere length restored by hTERT expression. The cell population would be expanded in culture and then cells would be reintroduced into the body to restore tissue and organ function. In this scheme, hTERT plays a role in autotransplantation. Telomerization could also be useful in allotransplantation and xenotransplantation by allowing expansion of cells with specific properties, such as stem cells or genetically modified cells.

Irrespective of the potential use of telomerization in future forms of cell therapy, cell transplantation offers an important method for determining whether cells have truly retained normal properties after telomerization. Our cell transplantation studies began as an outgrowth of studies of the behavior of adrenocortical cells grown in culture. The ability to grow human and animal cells outside the body has been developed over many years. These techniques are essential for cell transplantation; many procedures would be impossible if we did not know how to isolate specific cell types and then perform experiments that require cell culture, such as cloning cells or genetically modifying cells. Ideally, all of the requirements for normal function of the cells in vitro are known, but this is rarely achievable with current methods. Obviously, cell culture systems do not provide the complete three-dimensional structure, cell–cell interactions, blood supply, extracellular matrix, and paracrine influences that the cells have in vivo. Consequently, it is not really possible to test how normal cultured cells are unless they are challenged to form a functional tissue structure in vivo. Their ability to do so should be seen as the definitive test that they could still be considered to be normal despite whatever has happened to them in culture.

The methods of telomerization we have used have been applied to bovine and human adrenocortical cells *(18,19)* but should be generally applicable to many cell types. However, there is an ongoing controversy as to whether the only requirement for immortalization of some epithelial cell types is abrogation of telomere shortening (i.e., telomerization) or whether other blocks to indefinite growth must also be overcome. For example, for mammary epithelial cells, it was reported that expression of hTERT was insufficient by itself for immortalization of mammary epithelial cells *(20)*, but subsequent work by Ramirez et al. indicated that hTERT is the only genetic change needed when culture conditions are optimized for the specific cell type being studied *(21)*. For human adrenocortical cells, we also find that when culture conditions are optimized (as described below) ectopic expression of hTERT appears to be the only requirement for immortalization *(18,19)*.

2. Materials

2.1. Preparation of Primary Cells for Retroviral Infection

1. Type I-A collagenase (Sigma Chemical Co., St. Louis, MO).
2. DNAse I (Sigma).
3. Dulbecco's Eagle's medium.
4. Ham's F-12 medium.
5. Fetal bovine serum.
6. Heat-inactivated horse serum.
7. Recombinant FGF-2 (R & D Systems, Minneapolis, MN).
8. UltroSer G (Biosepra, Villeneuve-la-Garenne, France).
9. Collagen I-coated tissue culture dishes (Becton Dickinson Inc., Franklin Lakes, NJ).

2.2. Preparation of Retrovirus-Producing Cells and Retroviral Infection

1. Appropriate packaging cell line (e.g., PT67, Clontech Inc., Palo Alto, CA).
2. Plasmid pBabe-puro-hTERT *(22)*.
3. Transwell dishes (Corning Inc., Corning, NY).
4. Puromycin.

2.3. Telomerase Activity

1. Telomeric repeat amplification protocol (TRAP) assay kit (Intergen Co., Purchase, NY).
2. $[\alpha\text{-}^{32}P]dCTP$.

2.4. Cell Transplantation in Immunodeficient Mice

1. 3T3 cells that make a secreted form of FGF-1 *(23)*.
2. Mitomycin C (Sigma) or irradiator (Gammacell 1000 model C, AECL Industrial, Kanaka, Ontario).
3. ICR *scid* mice (Taconic, Germantown, NY).
4. Avertin (tribromoethanol/isoamyl alcohol; Sigma).
5. Polycarbonate tubing.
6. Chloroform.
7. Glass syringe (50 µL; Hamilton Company, Reno, NV).
8. Blunt 26-gage needle to fit glass syringe (needle length preferably approx 18 mm).
9. 6-0 nylon sutures.
10. Surgical staples.
11. Analgesics (acetaminophen, codeine).
12. Antibiotics (tetracycline, sulfamethoxazole, trimethoprim).

2.5. Cell Transplantation in Soluble Collagen

1. Soluble collagen (Cellagen; 0.2% pepsin-solubilized type I collagen, pH 7.4; ICN Pharmaceuticals, Costa Mesa, CA).
2. 29-Gage 1-mL disposable insulin injection syringe (Becton-Dickinson).

3. Methods

3.1. Telomerization of Adrenocortical Cells by Retroviral Transduction

3.1.1. Growth of Primary Cells for Retroviral Infection

Culture conditions for human adrenocortical cells have been optimized over several years, and these conditions are also ideal for bovine adrenocortical cells. Bovine adrenocortical cells are derived by enzymatic and mechanical dispersion from the adrenal cortex of 2-yr-old steers. Tissue fragments are dissociated to cell suspensions using enzymatic and mechanical dispersal (3-h incubation with 1 mg/mL collagenase and 0.1 mg/mL DNAse). Primary cell suspensions are stored frozen in liquid nitrogen. Frozen cells are thawed and replated in Dulbecco's Eagle's Medium/Ham's F-12 1:1 with 10% fetal bovine serum, 10% heat-inactivated horse serum, and 0.1 ng/mL recombinant FGF-2. This is supplemented with 1% v/v UltroSer G, a mixture of growth factors that greatly promotes adrenocortical cell growth in culture. The gas phase used is 90% N_2, 5% O_2, and 5% CO_2 *(24,25)*.

Human adrenal glands are obtained from kidney organ donors or from patients undergoing resection of the kidney for renal neoplasms. Tissue fragments are dissociated as described for bovine adrenocortical cells. Cells are plated and grown in culture as described for bovine adrenocortical cells, except that the cells grow better on collagen I-coated tissue culture dishes.

3.1.2. Preparation of Retrovirus-Producing Cells

For retroviral infection, a suitable packaging cell line is transfected with plasmid pBabe-puro-hTERT, encoding a retrovirus derived from mouse Moloney leukemia virus in which hTERT is expressed from the retroviral LTR (*see* **Note 1**) *(22)*. The plasmid is transfected by any convenient means into the PT67 mouse packaging cell line, which produces a virus type (10A1) that efficiently infects most human cells *(26)* and also bovine adrenocortical cells (our unpublished observations). Stably transfected cells are selected in 5 µg/mL puromycin. After nontransfected cells have been killed, the puromycin-resistant cell population is frozen in liquid nitrogen for future use.

3.1.3. Infection of Target Cells

For transduction of the target cells (adrenocortical cells), cells are plated in 10-cm Transwell dishes. The retrovirus-producing cells are plated in the insert, thus positioning them immediately above the target cells, separated by a polycarbonate membrane with 3-µm pores. The infection is allowed to take place over 4 d, followed by selection in puromycin. The concentration of puromycin should be optimized for the cell type being transduced; for human adrenocortical cells the effective concentration is 1 µg/mL. After antibiotic selection over 7–10 d, the cells are grown to a sufficient cell number for biochemical analysis and subsequent studies.

3.1.4. Telomerase Activity

The cells should be tested for acquisition of telomerase activity, by using the TRAP (telomerase repeat amplification protocol) assay (**Fig. 1**) *(27)*. Human and bovine adrenocortical cells, like most differentiated cell types, lack telomerase activity when grown over long periods in culture. They do, however, exhibit transitory telomerase activity when first placed in culture *(18)*. This activity rapidly diminishes to undetectable levels. After transduction with a retrovirus expressing hTERT, telomerase activity is very high and is sustained over long-term cell division (*see* **Note 2**). Stabilization of average telomere length may be monitored by measuring telomere restriction fragment length *(28)*.

3.2. Transplantation of Adrenocortical Cells in Severe Combined Immunodeficiency (Scid) Mice

3.2.1. Transplantation in the Subrenal Capsule Space

The preferred site for cell transplantation is beneath the kidney capsule (*see* **Note 3**) *(29)*. Adrenocortical cells are cotransplanted with 3T3-FGF cells (*see* **Note 4**). To render the 3T3-FGF cells incapable of further division after transplantation, they are treated with mitomycin C or are lethally irradiated. Cells are incubated at approx 20% confluence for 24 h with 2 µg/mL mitomycin C or are pelleted and exposed to 60 Gray radiation from a ^{137}Cs source.

The host animal must be an immunodeficient model, such as mice with the *scid* mutation (*see* **Note 5**). Animals (both males and females) at an age greater than 6 wk (~25 g body weight) are used. Mice are anesthetized with Avertin, which is made up, stored, and administered as described in detail elsewhere *(30)*. If necessary, bilateral adrenalectomy and cell transplantation may be performed in the same surgical procedure (*see* **Note 6**).

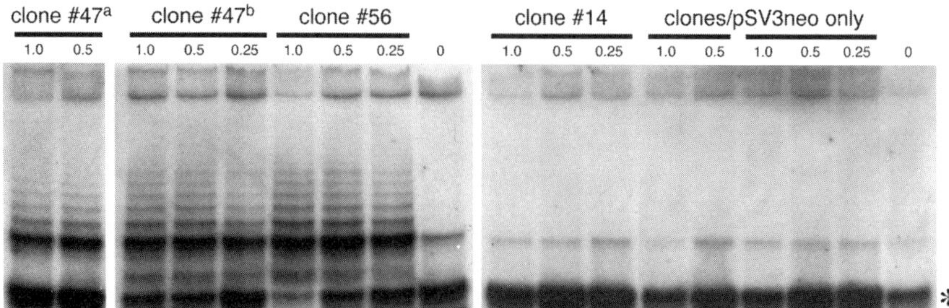

Fig. 1. Telomerase activity by TRAP assay in bovine adrenocortical cells expressing human telomerase reverse transcriptase (hTERT) and in control clones. For each clone, cell extracts at 1, 0.5, and 0.25 µg protein were assayed. 0 are reactions with cell lysis buffer only. Left, clones 47 and 56 were derived by transfection of primary bovine adrenocortical cells with pCI-neo-hTERT, pSV3neo, and pEGFP-N1 *(28)*. Clone 47 was assayed at two PDLs: (a) PDL 55 and (b) PDL 83. Clone 56 was assayed at PDL 25. Right, clone 14 is a nongenetically modified clone previously used for cell transplantation *(29)*. Two other clones were derived by transfection with pSV3neo and pEGFP-N1 without pCI-neo-hTERT. Amplification of the internal control fragment (marked with an asterisk) tests for possible inhibition of the PCR stage of the TRAP assay, which can produce false negatives. Clone 14 was assayed at PDL 30, and the two pSV3neo-only clones were assayed at PDL 25. Reproduced with permission from **ref.** *28*.

A longitudinal incision is made with fine scissors in the dorsal skin. A 1.5-cm incision in the lateral body wall is made to open the retroperitoneal space. In order to confine the transplanted adrenocortical cells within a defined space so that the growth, vascularization and function of the cells may be readily studied, a small polycarbonate cylinder is used to create a virtual space beneath the capsule into which the cells may be introduced *(29)*. A 1-mm length of polycarbonate tubing (2- to 3-mm internal diameter) is surface-polished by brief exposure to chloroform vapor (*see* **Note 7**). Cylinders are sterilized in 70% ethanol prior to use.

Using the incision in the body wall, the left kidney is exteriorized and a small transverse incision is made through the capsule on the ventral surface of the kidney near the inferior pole. Using one point of a fine forceps, a pocket is created under the capsule. A polycarbonate cylinder is pushed partially into the pocket under the capsule, filled with culture medium, and then introduced fully into the pocket so that the capsule on the top and the kidney parenchyma on the bottom form a sealed space. Adrenocortical cells are introduced into this space as follows. (2×10^6) adrenocortical cells are mixed with 4×10^5 3T3-FGF cells, pelleted by centrifugation, and kept on ice. For injection, the pellet is resuspended in medium, the total volume of the suspension being just greater than the volume of the pelletted cells. The cells are introduced into the subcapsular cylinder by a transrenal injection using a 50-µL syringe with a blunt 26-gage needle. After transplantation of the cells, the kidney is returned to the retroperitoneal space and bathed in approx 3 mL of phosphate-buffered saline. The body wall is closed with 6-0 nylon sutures and the skin is closed with surgical staples. Animals are maintained at 35°C ambient temperature until recovery from the anesthetic.

Postoperative care for the animals includes the administration of analgesics and antibiotics. Animals are given access to drinking water containing 1 mg/mL acetaminophen, 0.1 mg/mL codeine, 1 mg/mL tetracycline, 1 mg/mL sulfamethoxazole, and 0.1 mg/mL trimethoprim. The pH of the drinking water is adjusted to 6.05. If animals have been adrenalectomized, temporary administration of adrenocortical steroids may be necessary *(28,29)*. At 5 wk after transplantation, the transplanted cells have formed a mature, vascularized, functional tissue structure. **Figure 2** illustrates the histology of tissue formed from telomerized bovine adrenocortical cells. Note that there is no sign of neoplastic transformation; in particular, the rate of cell division is very low (**Fig. 3**). However, not all telomerized cell populations that we have tested can be successfully transplanted, success being defined as the formation of a functional tissue structure such as that illustrated in **Figs. 2** and **3** (*see* **Note 8**).

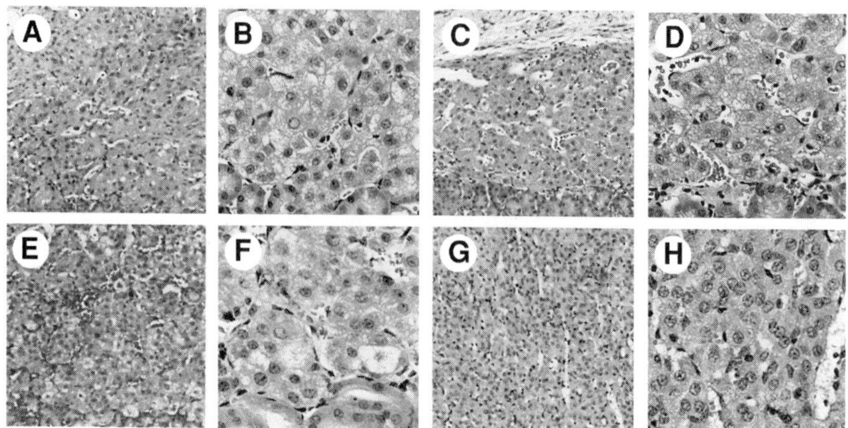

Fig. 2. Histological appearance of hematoxylin and eosin-stained tissues formed by transplantation of hTERT-modified bovine adrenocortical cell clones and normal adrenocortical tissue. (**A** and **B**) Examples of tissue formed by transplantation of clone 47. (**C** and **D**) Examples of tissue formed by transplantation of clone 56. (**E** and **F**) Tissue formed from transplanted cells of clone 14, which was not genetically modified. (**G** and **H**) Normal bovine adrenal cortex. (A–F) The mouse kidney may be seen below the transplant tissue. Magnifications ×250 (A, C, E, and G) and ×400 (B, D, F, and H). Reproduced with permission from **ref.** *28*.

3.2.2. Intradermal Cell Transplantation in Soluble Collagen

The subrenal capsule site is an excellent site for cell transplantation but is not necessarily ideal. Some of the disadvantages may be overcome by using intradermal cell transplantation (*see* **Note 9**).

In this procedure, the initial skin incision is made as for subrenal capsule transplantation (*see* **Subheading 3.2.1.**). A second skin incision is made at 90 degrees to the first, creating a triangular skin flap *(31)*. The inner surface of the skin flap is viewed under a dissecting microscope. A site next to larger blood vessels is chosen for intradermal injection. The cells to be transplanted are resuspended in 50 µL soluble collagen (0.2% pepsin-solubilized type I collagen, pH 7.4; Cellagen, ICN Pharmaceuticals, Costa Mesa, CA). The mixture of cells and collagen is kept on ice until use. The cell suspension is then placed in a disposable insulin syringe. Under the

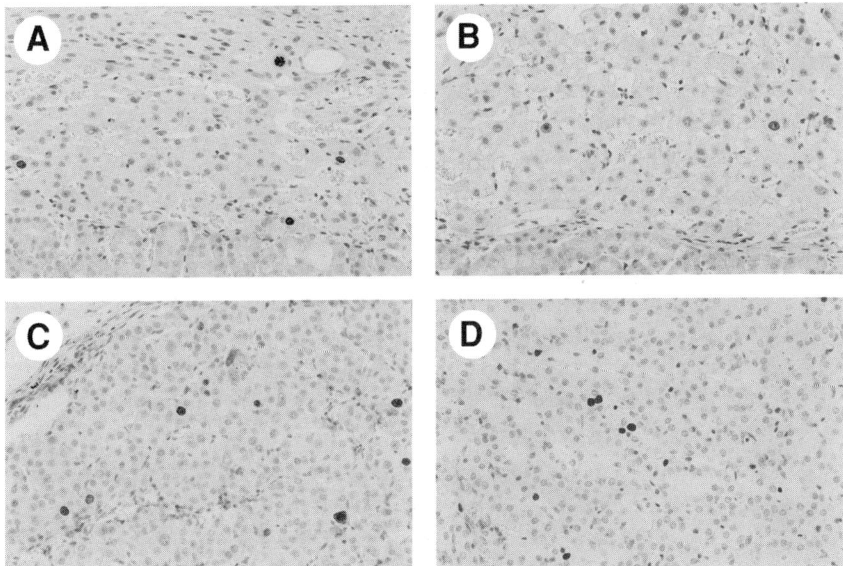

Fig. 3. Proliferation in tissues formed by transplantation of hTERT-modified bovine adrenocortical cell clones assessed by expression of Ki-67 antigen. (**A**) Tissue formed by transplantation of clone 47. (**B**) Tissue formed by transplantation of clone 56. (**C**) Tissue formed from transplanted cells of a clone that was not genetically modified (clone 14). (**D**) Normal bovine adrenal cortex. Magnification ×300. Reproduced with permission from **ref.** *28*.

microscope, the cell suspension is injected intradermally. To minimize leakage from injection site, the needle is not withdrawn until a suture (6-0 nylon) is placed to close the needle track. Postoperative care for the animals is the same as for subrenal capsule cell transplantation.

4. Notes

1. The telomerized bovine adrenocortical cells used in our published cell transplantation experiments *(28)* were actually derived by transfection rather than retroviral transduction. At the time these telomerized clones were created, we did not yet have access to an hTERT-expressing retrovirus. The telomerized clones were created

by cotransfection of cells with pCI-neo-hTERT, a plasmid in which hTERT is expressed from the CMV promoter, and pSV3neo, a plasmid containing the early region of SV40 virus that expresses both large T and small t antigens. We were forced to use cotransfection because we found that stably transfected clones were not formed when the hTERT plasmid, or any other plasmids, were used without the cotransfection of pSV3neo. We theorize that this dependence on the inclusion of a plasmid-encoding SV40 T antigen might result from direct adverse effects of the transfection process itself in primary cells. Transfection involves the introduction of DNA fragments into cells, which could cause a permanent cell cycle block similar to replicative senescence, that is, a reaction to the presence of DNA double-strand breaks. In support of this hypothesis, transfection has been shown to increase p21, an indicator of a p53-dependent block (J. R. Smith and S. F. Venable, unpublished observations). T antigen is known to prevent the normally growth-inhibitory effects of p53 by binding to and inactivating it *(32)*. Thus, it is possible the expression of T antigen (from plasmid pSV3neo) at the time of transfection enables cells to divide rather than be subjected to cell cycle block. Once past this point, further action of T antigen might be unnecessary, because the unintegrated plasmid DNA would have been degraded. However, one must note that the telomerized clones of cells we used to form functional tissue after cell transplantation were formed by coexpression of hTERT and SV40 T antigen. What role T antigen may play, other than preventing the immediate cell cycle block caused by the transfection procedure itself, is not clear. In culture, T antigen is expressed in the clones used (clones 47 and 56), although at considerably lower levels than in cells in which T antigen is expressed from a retroviral LTR *(33)*. This is presumably because the promoter (the SV40 promoter) is not as strong as some other viral promoters. Moreover, in the tissues formed by transplantation of these cells we found an almost undetectable level of expression of T antigen by immunocytochemistry, probably because the SV40 promoter is mostly inactive in tissues in vivo *(34)*. In contrast, when cells that expressed T antigen from an LTR were transplanted, immunocytochemistry showed high levels of expression *(33)*. These cells also showed a very high rate of cell proliferation, whereas the rate of cell proliferation in tissues formed from clones 47 and 56 was no higher than that in tissue formed from normal nonmodified cells *(28)*.

Subsequently, we introduced hTERT alone into bovine adrenocortical cells using retroviral techniques, as described previously. Despite the high level of telomerase activity in these cells, and the probable immortalization of these cells by hTERT, cells did not form tissue as efficiently as cells that were telomerized by cotransfection of pCI-neo-hTERT and pSV3neo; they were not able to form tissue when grown over long periods of time (unpublished observations). This may indicate that SV40 T antigen expressed at low levels from the SV40 promoter may increase and/or stabilize differentiated functions in adrenocortical cells, in agreement with earlier work in cell culture *(35,36)*. More study is needed to establish whether low levels of T antigen might generally have desirable effects in telomerization of differentiated cell types.

2. To maintain telomerase activity and thereby to avoid replicative senescence, TERT must continue to be expressed from the integrated retroviral DNA. It is possible that the retroviral promoter (in this case the LTR) may cease to function in the long term, due to changes in DNA methylation, histone acetylation, or other mechanisms *(37)*. However, those cells within the population that continue to express TERT will take over the population at the expense of those that do not; thus, expression of TERT acts as an agent of positive selection.

3. Transplantation beneath the kidney capsule has been used for many years to support the survival and growth of embryonic organs and some adult tissues, including the adrenal cortex *(38–40)*. Some tumors that do not grow when transplanted subcutaneously can often be propagated when placed beneath the kidney capsule. There is no evidence that this is an immune privileged site; the important feature of this site is probably that the extracellular fluid pressure is relatively high, giving the transplanted cells immediate access to adequate nutrients and oxygen before vascularization occurs. These features are probably lacking when cells are transplanted beneath the skin, except when care is taken to protect the cells from fluid loss (*see* **Subheading 3.2.2.**).

4. Over the first few days after transplantation of a cell suspension the cells must form cell–cell contacts and cooperate with host endothelial cells to create a functional vasculature. The support of angiogenesis by normal adult cells and tissues has not been as well studied as the use of angiogenesis by cancer cells to overcome limitations on

tumor size. Because we were unsure of the capacity of normal cells to support angiogenesis, we began our cell transplantation studies by cotransplanting with the adrenocortical cells a small number of 3T3 cells which make a secreted form of FGF-1 *(23,29)*. The cells are treated before cell transplantation to make them incapable of cell division. They do not form an obvious part of the transplant tissue; they may resemble pericytes. However, subsequently we showed that bovine adrenocortical cells are capable of tissue formation without assistance for angiogenesis *(41)*. When they are transplanted without 3T3 cells the tissue that forms at 36 d shows areas of necrosis, even though the rest of the tissue is apparently healthy and functional. The necrotic area eventually resolves with the formation of healthy tissue about 70 d after introduction of the cells into the animal. When FGF-secreting 3T3 cells are included, the transplant tissue does not show a phase where necrosis is present. The best current hypothesis for the action of the cotransplanted 3T3-FGF cells is that they act in a supportive role, perhaps in a manner similar to feeder layers in culture. More study is needed to define their role.

5. Observing the properties of telomerized cells in host animals requires xenotransplantation, the introduction of human or bovine cells into a mouse. There have been no comprehensive studies of what specific immunodeficiencies are required for xenotransplantation of normal and telomerized cells; reports of successful xenotransplantation have most often used *scid* mice *(42–44)*. However, newer immunodeficient mouse models, such as the RAG2/cytokine receptor gamma subunit double knockout, are very promising *(45)*. When there are failures of cells to perform normally in cell transplantation in *scid* mice (e.g., telomerized cells grown over long periods in culture; **ref. 28**) we do not yet know whether the failure could be corrected by the use of different immunodeficient mice. One reason for testing other models is the lower natural killer cell activity in comparison to the *scid* mouse; it is possible that as telomerized and other cells are grown over long periods they acquire changes that render them susceptible to being recognized as foreign by natural killer cells *(46)*.

6. Adrenalectomy is used to demonstrate that animals survive the normally lethal procedure of adrenalectomy when they receive transplants of telomerized adrenocortical cells, thus demonstrating that these cells can successfully replace the essential functions of the ani-

mals' adrenal glands *(28)*. However, adrenalectomy is not an essential part of the cell transplantation procedure.

7. Polycarbonate cylinders are polished by brief exposure to chloroform vapor or by brief immersion in chloroform. Polishing removes rough surfaces that can provoke a foreign body reaction. Please note that, erroneously, cylinders were originally described as being polished with dichloromethane *(29)*.

8. Much of the process of the formation of tissue from telomerized cells by cell transplantation is currently more art than science. We have no ready explanations for the failures of some cell populations to form tissue. For example, we noted that telomerized clone 47 of bovine adrenocortical cells became progressively less efficient at tissue formation after long periods of growth in culture *(28)*. When cells are grown for long periods of time in culture there are likely to be changes in gene expression that may adversely affect the ability of cells to form tissue. Such changes may not necessarily be reversed by telomerization. However, at least in skin fibroblasts, telomerization prevents changes in gene expression associated with replicative senescence and permits the production of a normal skin equivalent after transplantation *(16)*.

9. For the reasons described above, we have primarily used the subrenal capsule site for cell transplantation, but the use of other sites in the body is clearly feasible *(47,48)*. The ideal site would permit rapid establishment and vascularization of the transplant combined with easy access for repeated observations (e.g., by cutaneous windows; **ref. *49***). The subrenal capsule has consistently given the best results, despite success in other sites of cell transplantation. However, it is not conveniently accessible for repeated interventions or observations, and there is a limitation on the number of cells that can be introduced without risking damage to the kidney or leakage of the cell suspension. Cell transplantation beneath the skin has the advantages of ready accessibility and no limitation on the number of cells that can be implanted in one place. In the mouse, which is commonly the host for cell transplantation experiments because of the availability of immunodeficient models, the subcutaneous site is not hospitable for cell transplantation. Moreover, the anatomy of the skin and underlying body wall does not permit cells to be readily confined to one site. However, intradermal, as opposed

to subcutaneous, cell transplantation is advantageous for tumor cell growth in the mouse *(31,50,51)*. For normal cells, we have shown that when intradermal cell transplantation is used for cells suspended in soluble collagen, a discrete tissue structure is produced in a defined location *(52)*.

Acknowledgments

Work from the author's laboratory was supported by grants from the National Institute on Aging (AG 12287 and AG 20752) and by a Senior Scholar Award from the Ellison Medical Foundation.

References

1. Hayflick, L. (1980) Cell aging. *Annu. Rev. Gerontol. Geriatr.* **1,** 26–67.
2. Harley, C. B., Futcher, A. B., and Greider, C. W. (1990) Telomeres shorten during ageing of human fibroblasts. *Nature* **345,** 458–460.
3. Allsopp, R. C., Vaziri, H., Patterson, C., Goldstein, S., Younglai, E. V., Futcher, A. B., et al. (1992) Telomere length predicts replicative capacity of human fibroblasts. *Proc. Natl. Acad. Sci. USA* **89,** 10,114–10,118.
4. Greider, C. W. (1990) Telomeres, telomerase and senescence. *Bioessays* **12,** 363–369.
5. Harley, C. B. (1991) Telomere loss: Mitotic clock or genetic time bomb? *Mutat. Res.* **256,** 271–282.
6. Lingner, J., Hughes, T. R., Shevchenko, A., Mann, M., Lundblad, V., and Cech, T. R. (1997) Reverse transcriptase motifs in the catalytic subunit of telomerase. *Science* **276,** 561–567.
7. Meyerson, M., Counter, C. M., Eaton, E. N., Ellisen, L. W., Steiner, P., Caddle, S. D., et al. (1997) hEST2, the putative human telomerase catalytic subunit gene, is up-regulated in tumor cells and during immortalization. *Cell* **90,** 785–795.
8. Hornsby, P. J. Cell proliferation in mammalian aging, in *Handbook of the Biology of Aging* 5th ed (Masoro, E. J., and Austad, S. N., eds.), Academic Press, San Diego, 2001, pp. 207–266.
9. Noda, A., Ning, Y., Venable, S. F., Pereira-Smith, O. M., and Smith, J. R. (1994) Cloning of senescent cell-derived inhibitors of DNA synthesis using an expression screen. *Exp. Cell Res.* **211,** 90–98.

10. Smith, J. R. and Pereira-Smith, O. M. (1996) Replicative senescence: Implications for in vivo aging and tumor suppression. *Science* **273,** 63–67.

11. Bodnar, A. G., Ouellette, M., Frolkis, M., Holt, S. E., Chiu, C. P., Morin, G. B., et al. (1998) Extension of life-span by introduction of telomerase into normal human cells. *Science* **279,** 349–352.

12. Jiang, X. R., Jimenez, G., Chang, E., Frolkis, M., Kusler, B., Sage, M., et al. (1999) Telomerase expression in human somatic cells does not induce changes associated with a transformed phenotype. *Nat. Genet.* **21,** 111–114.

13. Morales, C. P., Holt, S. E., Ouellette, M., Kaur, K. J., Yan, Y., Wilson, K. S., et al. (1999) Absence of cancer-associated changes in human fibroblasts immortalized with telomerase. *Nat. Genet.* **21,** 115–118.

14. Weinberg, R. A. (1998) Telomeres. Bumps on the road to immortality. *Nature* **396,** 23–24.

15. de Lange, T. and DePinho, R. A. (1999) Unlimited mileage from telomerase? *Science* **283,** 947–949.

16. Funk, W. D., Wang, C. K., Shelton, D. N., Harley, C. B., Pagon, G. D., and Hoeffler, W. K. (2000) Telomerase expression restores dermal integrity to in vitro-aged fibroblasts in a reconstituted skin model. *Exp. Cell Res.* **258,** 270–278.

17. Shay, J. W. and Wright, W. E. (2000) The use of telomerized cells for tissue engineering. *Nat. Biotechnol.* **18,** 22–23.

18. Suwa, T., Yang, L., and Hornsby, P. J. (2001) Telomerase activity in primary cultures of normal adrenocortical cells. *J. Endocrinol.* **170,** 677–684.

19. Yang, L., Suwa, T., Wright, W. E., Shay, J. W., and Hornsby, P. J. (2001) Telomere shortening and decline in replicative potential as a function of donor age in human adrenocortical cells. *Mech. Age. Dev.* **122,** 1685–1694.

20. Kiyono, T., Foster, S. A., Koop, J. I., McDougall, J. K., Galloway, D. A., and Klingelhutz, A. J. (1998) Both Rb/p16INK4a inactivation and telomerase activity are required to immortalize human epithelial cells. *Nature* **396,** 84–8.

21. Ramirez, R. D., Passons, C., Rohde, J., Morales, C. P., Herbert, B.-S., Shay, J. W., et al. (2001) Putative telomere-independent mechanisms of replicative aging reflect inadequate growth conditions. *Genes Dev.* **15,** 398–403.

22. Kim, S. H., Kaminker, P., and Campisi, J. (1999) TIN2, a new regulator of telomere length in human cells. *Nat. Genet.* **23,** 405–412.

23. Forough, R., Xi, Z., MacPhee, M., Friedman, S., Engleka, K. A., Sayers, T., et al. (1993) Differential transforming abilities of non-secreted and secreted forms of human fibroblast growth factor-1. *J. Biol. Chem.* **268,** 2960–2968.

24. Hornsby, P. J. and McAllister, J. M. Culturing steroidogenic cells, in *Methods in Enzymology*, Vol. 206 (Waterman, M. R., and Johnson, E. F., eds.), Academic, San Diego, 1991, pp. 371–380.

25. Hornsby, P. J. Culturing steroidogenic cells, in *Cell and Tissue Culture: Laboratory Procedures*, 17B (Griffiths, J. B., Doyle, A., and Newell, D. G., eds.), Wiley, Chichester, UK, 1994, pp. 7.1–7.9.

26. Miller, A. D. and Chen, F. (1996) Retrovirus packaging cells based on 10A1 murine leukemia virus for production of vectors that use multiple receptors for cell entry. *J. Virol.* **70,** 5564–5571.

27. Kim, N. W., Piatyszek, M. A., Prowse, K. R., Harley, C. B., West, M. D., Ho, P. L., et al. (1994) Specific association of human telomerase activity with immortal cells and cancer. *Science* **266,** 2011–2015.

28. Thomas, M., Yang, L., and Hornsby, P. J. (2000) Formation of normal functional tissue from transplanted adrenocortical cells expressing telomerase reverse transcriptase. *Nat. Biotechnol.* **18,** 39–42.

29. Thomas, M., Northrup, S. R., and Hornsby, P. J. (1997) Adrenocortical tissue formed by transplantation of normal clones of bovine adrenocortical cells in *scid* mice replaces the essential functions of the animals' adrenal glands. *Nat. Med.* **3,** 978–983.

30. Papaioannou, V. E. and Fox, J. G. (1993) Efficacy of tribromoethanol anesthesia in mice. *Lab. Anim. Sci.* **43,** 189–192.

31. Runkel, S., Hunter, N., and Milas, L. (1991) An intradermal assay for quantification and kinetics studies of tumor angiogenesis in mice. *Radiat. Res.* **126,** 237–243.

32. Butel, J. S. (1986) SV40 large T-antigen: Dual oncogene. *Cancer Surv.* **5,** 343–365.

33. Thomas, M., Suwa, T., Yang, L., Zhao, L., Hawks, C. L., and Hornsby, P. J. (2002) Cooperation of hTERT, SV40 T antigen and oncogenic Ras in tumorigenesis: A cell transplantation model using bovine adrenocortical cells. *Neoplasia* **4,** 493–500.

34. Stewart, C. L., Schuetze, S., Vanek, M., and Wagner, E. F. (1987) Expression of retroviral vectors in transgenic mice obtained by embryo infection. *EMBO J.* **6,** 383–388.

35. Cheng, C. Y., Ryan, R. F., Vo, T. P., and Hornsby, P. J. (1989) Cellular senescence involves stochastic processes causing loss of expression of differentiated function genes: Transfection with SV40 as a means for dissociating effects of senescence on growth and on differentiated function gene expression. *Exp. Cell Res.* **180,** 49–62.

36. Cheng, C. Y., Flasch, M. V., and Hornsby, P. J. (1992) Expression of 17α-hydroxylase and 3β-hydroxysteroid dehydrogenase in fetal human adrenocortical cells transfected with SV40 T antigen. *J. Mol. Endocrinol.* **9,** 7–17.

37. McInerney, J. M., Nawrocki, J. R., and Lowrey, C. H. (2000) Long-term silencing of retroviral vectors is resistant to reversal by trichostatin A and 5-azacytidine. *Gene Ther.* **7,** 653–663.

38. Everett, N. B. (1949) Autoplastic and homoplastic transplants of the rat adrenal cortex and medulla to the kidney. *Anat. Rec.* **103,** 335–347.

39. Scheumann, G. F.W., Hiller, W. F., Schroder, S., Schurmeyer, T., Klempnauer, J., and Dralle, H. (1989) Adrenal cortex transplantation after bilateral total adrenalectomy in the rat. *Henry Ford Hosp. Med. J.* **37,** 154–156.

40. Scheumann, G. F. W., Heitmann, P., Teebken, O. E., Mossinger, E., Mellon, S. H., and Pichlmayr, R. (1996) Enzymatic properties of transplanted glomerulosa cells. *World J. Surg.* **20,** 933–939.

41. Thomas, M. and Hornsby, P. J. (1999) Transplantation of primary bovine adrenocortical cells into *scid* mice. *Mol. Cell. Endocrinol.* **153,** 125–136.

42. Bosma, M. J., Phillips, R. A., and Schuler, W. *The scid Mouse: Characterization and Potential Uses.* Springer-Verlag, Berlin, 1989.

43. Martin, A., Valentine, M., Unger, P., Lichtenstein, C., Schwartz, A. E., Friedman, E. W., et al. (1993) Preservation of functioning human thyroid organoids in the scid mouse: 1. System characterization. *J. Clin. Endocrinol. Metab.* **77,** 305–310.

44. Volpe, R., Akasu, F., Morita, T., Yoshikawa, N., Resetkova, E., Arreaza, G., et al. (1993) New animal models for human autoimmune thyroid disease. Xenografts of human thyroid tissue in severe combined immunodeficient (SCID) and nude mice. *Horm. Metab. Res.* **25,** 623–627.

45. Cooper, R. N., Irintchev, A., Di Santo, J. P., Zweyer, M., Morgan, J. E., Partridge, T. A., et al. (2001) A new immunodeficient mouse model for human myoblast transplantation. *Hum. Gene Ther.* **12,** 823–831.

46. Seebach, J. D. and Waneck, G. L. (1997) Natural killer cells in xenotransplantation. *Xenotransplantation* **4,** 201–211.
47. Hornsby, P. J., Thomas, M., Northrup, S. R., Popnikolov, N. P., Wang, X., Tunstead, J. R., et al. (1998) Cell transplantation: A tool to study adrenocortical cell biology, physiology, and senescence. *Endocr. Res.* **24,** 909–918.
48. Ciancio, S. J., King, S. R., Suwa, T., Thomas, M., Yang, L., Zhang, H., et al. (2000) Transplantation of normal and genetically modified adrenocortical cells. *Endocr. Res.* **26,** 931–940.
49. Ciancio, S. J., Coburn, M., and Hornsby, P. J. (2000) A cutaneous window allowing in vivo observations of organs and angiogenesis in the mouse. *J. Surg. Res.* **92,** 228–232.
50. Dunham, W. B. and Waymouth, C. (1976) Intradermal transplantation in mice of small numbers of sarcoma cells followed by tumor growth and regression. *Cancer. Res.* **36,** 189–193.
51. Cornil, I., Man, S., Fernandez, B., and Kerbel, R. S. (1989) Enhanced tumorigenicity, melanogenesis, and metastases of a human malignant melanoma after subdermal implantation in nude mice. *J. Natl. Cancer Inst.* **81,** 938–944.
52. Zhang, H. and Hornsby, P. J. (2002) Intradermal cell transplantation in soluble collagen. *Cell Transplantation* **11,** 139–145.

9

Chromosome-Based Vectors for Mammalian Cells

An Overview

Han N. Lim and Christine J. Farr

1. Introduction

The functional organization of the eukaryotic chromosome was first elucidated at a molecular level in the budding yeast, *Saccharomyces cerevisiae*, providing the basis for the successful creation of yeast artificial chromosomes (YACs) *(1)*. The structures that confer chromosome function have been far more difficult to determine in multicellular eukaryotes, both because of their greater complexity and size. Over the last decade, various strategies have been developed for creating engineered human/mammalian chromosomes. These fall into two broad categories: the use of naked DNA containing sequences capable of *de novo* chromosome formation (the "bottom-up" approach) or the manipulation and modification of existing chromosomes (the "top-down" approach). In this review, we will refer to chromosomes formed from naked input DNA as artificial chromosomes (ACs) and those produced from existing chromosomes as mini- or derivative-chromosomes (depending on their final size).

From: *Methods in Molecular Biology, Vol. 240:*
Mammalian Artificial Chromosomes: Methods and Protocols
Edited by: V. Sgaramella and S. Eridani © Humana Press Inc., Totowa, NJ

As well as helping provide a better understanding of normal chromosome function, the potential of chromosome-based gene transfer vectors in transgenic animals and as human gene therapy agents has raised their profile in the wider community. This review aims to describe the basic functions of vertebrate chromosomes that an engineered chromosome needs to successfully recreate, the strategies for their construction, and their current and potential uses. Although the review will focus on human chromosomes, it should be emphasized that the possibility of species differences in chromosome structure and function requirements has yet to be fully explored, despite the long history of the somatic cell hybrid.

2. What Features Do Human Chromosomes Possess?

Human chromosomes share the same fundamental components of all eukaryotic chromosomes in that they have a centromere, two telomeres, and multiple replication origins. Telomeres are the protein–DNA structures that cap and protect linear chromosome ends *(2)*. The DNA component of all vertebrate telomeres is an array of $ds(TTAGGG)_n$. The introduction of just a few hundred base pairs of $ds(TTAGGG)_n$ repeat is sufficient to induce telomere formation in many vertebrate cell lines *(3–5)*. In contrast to the strict sequence requirements for telomeres the extent to which origin activity is dependent on specific DNA elements is still far from clear *(6)*; however, origins occur frequently in the human genome and our ignorance of origin DNA requirements has not posed a problem to the development of artificial chromosomes. However, the DNA requirements underlying the centromere/kinetochore structure are also somewhat equivocal and this is a critical issue that cannot be side stepped.

Microscopic visualization of the human metaphase chromosome simply reveals the centromere as a region of constriction located within heterochromatin. However, the centromere is perhaps the most complex of chromosome structures with a multitude of functions *(7)*. The centromere consists of an underlying DNA component upon which a complex of proteins known as the kinetochore assembles. The kinetochore provides the interface between the underlying chromatin and the spindle microtubules. In addition, the flanking chromatin may play key roles in sister chromatid cohesion and in providing a heterochromatic environment. Mammalian centromeric DNA sequences show little evolutionary conservation although a common feature appears to be the presence of vast expanses of generally (A + T)-rich satellite DNAs. In humans and other primates this DNA is α-satellite (also known as alphoid DNA), which is composed of a 171 bp monomer that is organized into tandem arrays. These arrays may have a higher order structure, which is itself repeated and can extend for many megabases (Mb). α-satellite DNAs show variation in both their sequence and in the higher order repeat organization between nonhomologous chromosomes, while the size of the arrays is highly polymorphic *(8–10)*. Despite this extensive variation, α-satellite sequences are recognizable on all normal human chromosomes and one study that examined over 17,000 normal human chromosomes failed to identify any without it *(10)*. The consistent presence of α-satellite at human centromeres implies a role in centromere function. Nevertheless, the existence of two types of variant centromere—dicentrics and analphoid neocentromeres—has revealed that primary DNA sequence is not a strict determinant of where the kinetochore assembles and suggests the involvement of, as yet undefined, epigenetic factors *(11,12)*.

3. Strategies for Engineering Chromosomes

3.1. De Novo *Artificial Chromosome Formation*

The first-generation human ACs were constructed using the principles of YAC assembly. Therefore, it was expected that a human AC would require telomere sequences, centromere sequences, and DNA that would provide origins of replication. The first human ACs to be described were constructed from unlinked components comprising synthetic α-satellite DNA, short arrays of ds(TTAGGG)$_n$, random genomic DNA fragments, and a positive selectable marker gene. This naked DNA mixture was transfected by lipofection (either with or without a preligation step) into human HT1080 tissue culture cells. A fluorescence *in situ* hybridization screen of drug-resistant transfectants recovered revealed some with additional minichromosomes. In most cases, these minichromosomes were found to be derived from existing HT1080 chromosomes as a result of telomere-induced fragmentation events. However, in one cell line, a *de novo* human AC, composed of the input 17 α-satellite DNA, was present *(13)*. After this study several groups independently created mitotically stable *de novo* human ACs using preassembled YAC- or P1 artificial chromosomes (PAC)-based vectors containing α-satellite DNA *(14–19)*.

3.1.1. What Have These Studies Revealed About De Novo Human AC Requirements?

It is apparent from the *de novo* human AC construction studies that large fragments of mammalian DNA, including a PAC/YAC backbone and α-satellite sequences, are replication competent. Another chromosomal element that does not appear to be necessary for AC formation or segregation is the telomere. In most studies, where it has been examined, the evidence suggests that the *de novo* chromosomes are circular in structure, not linear, irrespective of the presence or absence of ds(TTAGGG)$_n$ repeats in the input DNA. A detailed study examining the effect of varying the status of an input alphoid-containing PAC from circular to linear, capped or uncapped by ds(TTAGGG)$_n$ repeats *(17)* found that the presence of ds(TTAGGG)$_n$ repeats increased the efficiency of *de novo* chromosome formation. However, the ds(TTAGGG)$_n$ repeat had the same effect irrespective of its orientation and the structure of the *de novo* minichromosomes appeared to be circular. The authors concluded that the presence of ds(TTAGGG)$_n$ repeats in any orientation simply served to protect the ends of the input DNA molecules from the action of nucleases and recombinases (presumably through the binding of telomere-associated factors like TRF1 and TRF2) rather than having a direct role in *de novo* telomere formation, which would be required for the recovery of a structurally and mitotically stable linear minichromosome.

The average length of the input α-satellite sequences in the *de novo* assembly studies has been in the order of approx 100–200 kb. However, the estimated size of the *de novo* structures generated has been 1 to 10 Mb. Although the nature of the cell-mediated assembly process is unknown it is assumed that amplification and/or concatemerization of the input DNA molecules must occur to generate chromosome structures 10–100-fold the size of the input DNA molecules. Whether the final size recovered reflects the requirements of a stable *de novo* circular mammalian chromosome or whether it is simply a consequence of the assembly process is not clear.

Although α-satellite DNA derived from the main array on chromosomes 17, 21, and the X have been shown to form *de novo* ACs *(13,14,17–19)*, α-satellite sequences from the pericentromeric region of human chromosome 21 and from the human Y (DYZ3) appear very inefficient at forming *de novo* minichromosomes *(14,16,19)*. An intriguing difference between the *de novo* centromere-competent and -incompetent input alphoid DNAs is that the latter lack CENP-B boxes. It appears that for efficient *de novo* centromere formation (and hence AC recovery) a relatively homogeneous stretch of α-satellite DNA of sufficient length, exhibiting higher order repeat structure and containing CENP-B boxes is required *(20,21)*. The precise role of the CENP-B box in this process is unclear: the CENP-B protein itself is highly conserved in eukaryotes and yet the creation of mice lacking CENP-B had no detectable effect on chromosome behavior *(22–25)*. Moreover, binding sites for the CENP-B protein are naturally absent from the human Y centromere, from the centromeres of the primate African green monkey *(26,27)* and from human neocentromeres *(28,29)*, suggesting that CENP-B binding is not required for centromere function. Whether the presence of regularly distributed CENP-B binding sites in α-satellite DNA has a direct role in promoting *de novo* centromere/minichromosome formation or reflects an indirect requirement for a particular chromatin structure that DNA with regularly spaced CENP-B boxes more readily adopts has yet to be determined *(20,30)*.

As yet most of the *de novo* AC studies have been undertaken in the human tissue culture cell line HT1080. This adherent cell line is derived from a fibrosarcoma, has a recognizable and fairly stable human karyotype and is readily transfectable. The requirements of the recipient cells used in *de novo* AC experiments have not yet been established although successful AC generation in the human cell line HeLa has recently been reported *(31)*. Moreover, species differences/ similarities have still to be investigated. For example, is it possible to generate a *de novo* AC in human cells using mouse centromeric minor satellite DNA as the input and vice versa?

3.2. Manipulation of Existing Chromosomes

3.2.1. Chromosome Fragmentation Using Cloned Telomeric DNA

The basis of the chromosome fragmentation approach is the removal of all sequences that are not essential to the stable self-propagation of a chromosome. To avoid effects on cell viability the human chromosome manipulated is maintained as a nonessential chromosome in a somatic cell hybrid. The size of the human chromosome can be reduced by the introduction of short stretches (a few hundred base pairs) of the vertebrate telomere repeat ds(TTAGGG)$_n$. The ds(TTAGGG)$_n$ repeat is cloned into a plasmid carrying a suitable positive selectable marker gene. The plasmid DNA is introduced into the recipient cells by transfection in the linear form, having been restricted at a unique site immediately distal to the repeat *(32)*. The orientation of the repeat is such that on the linear ingoing DNA molecule it mimics the natural chromosome end. Upon integration the repeat can seed *de novo* telomere formation. If the *de novo* telomere is associated with an acentric chromosome arm it will be lost from the cell. If it caps a centromere-bearing chromosome fragment a stable truncation product will be recovered *(3–5,33)*. Irradiation has been used as an alternative to telomere-associated chromosome fragmentation for inducing chromosomal breaks, although the experimenter has less control over the outcome *(34–36)*.

Early attempts to generate human minichromosomes using telomere-associated chromosome fragmentation relied on random integration, since homologous recombination generally occurs at a very low frequency in mammalian cell lines. More recently human chromosomes have been transferred, using cell fusion techniques, into the recombination-proficient chicken cell line DT40 *(37–41)* making it practical to target telomere-seeding events and generate products of defined structure. Once manipulation of the target chromosome is completed in the DT40 cell line, it can then be shuttled back into mammalian cell lines for further characterization using cell fusion.

Telomere-associated chromosome fragmentation techniques have generated minichromosomes derived from the human X and Y chromosomes, as well as from the mardel (10) neocentromere-carrying marker chromosome *(42–47)*. The Y and X chromosome derivatives are linear, structurally stable and both minichromosome series have been systematically studied to determine mitotic stability in the absence of drug selection. The smallest reported Y-derived minichromosome is approx 1.8 Mb and retains only approx 100 kb of α-satellite DNA *(46)*. The human X centromere-based minichromosomes range in overall size from several megabases down to <1 Mb, with some retaining <200 kb of the major X α-satellite locus DXZ1 *(43,44,48)*. The X- and Y-derived minichromosomes demonstrate a high level of mitotic stability in the absence of selection in a variety of vertebrate tissue culture cell lines *(44,46)*; however, evidence for species differences in centromere requirements has been presented *(42,45)*. The observations on human X- and Y-derived minichromosomes, together with Q-FISH and PFGE data on some naturally occurring human chromosomes, suggest that for a pre-existing centromere approx 100 kb of alphoid DNA is sufficient for continued maintenance and function (irrespective of whether CENP-B boxes are present). The neocentromere-based minichromosomes that have been reported range in size from approx 0.7 to 1.8 Mb, with both circular and linear derivatives recovered

(47). These mammalian minichromosomes, which lack typical centromeric repeat DNA, were found to be both structurally stable and to show low loss rates in HT1080 cells.

3.2.2. Other Manipulation Techniques

Various other techniques have been used to alter endogenous chromosomes. One involves the integration of exogenous DNA into the pericentromeric regions of a host human or mouse chromosome. Amplification of DNA at the integration site can result in breakage events and the generation of stable *de novo* satellite DNA-based chromosomes in the 60–400 Mb size range *(49,50)*. Another strategy has been to manipulate naturally occurring marker chromosomes to reduce their size using irradiation, or to introduce selectable cassettes and genes of interest through random or targeted integration *(34,36,51,52)*.

3.3. Transfer of Engineered Chromosomes into Cells

The relatively large size of *de novo* ACs and minichromosomes limits the techniques that can be used to transfer them into cells. In the *de novo* centromere experiments, the components have been introduced as naked DNA, using transfection techniques such as lipofection. Improvements in the efficiency of uptake of large DNA molecules may be obtained by targeting the alphoid DNA constructs to specific cell surface receptors, by using condensing reagents (such as polyethylenimine), or through using viral packaging. Thus, introduction of large naked DNA molecules with a reasonable level of efficiency, although not trivial, does seem feasible. The major limitation with *de novo* assembly is that the experimenter has little influence over the outcome of the cell-mediated process and this may be complicated by the input DNA inserting randomly into the recipient genome causing insertional mutagenesis or unwanted fragmentation events.

To transfer existing chromosomes as chromatin cell fusion (whole cell or microcell) is generally the technique of choice. However, fusion is an extremely inefficient process, fragmentation may occur and unwanted chromosomes from the donor cell may also be transmitted. Recently a protocol has been devised for purifying large satellite DNA-based chromosomes as chromatin using a fluorescence-activated cell sorter. This creates a soup of purified ACs with their associated proteins still attached, which can be transferred into cultured mammalian cells using either cationic dendrimers, lipids, or by microinjection into mouse pronuclei *(53,54)*. Purification of these satellite DNA-based chromosomes away from the host chromosomes by fluorescence-activated cell sorting is possible because of their large size and biased base composition. Flow sorting of small minichromosomes based purely on size would likely produce heavy contamination from fragmented chromosomal debris and as yet successful strategies have not been reported.

3.4. Transgene Expression

A potential problem with chromosome-based vector applications is centromere-associated gene silencing, which has been described in yeast, flies, and mice. Genes have been integrated into engineered chromosomes both through the manipulation of existing minichromosomes in cultured cells or through incorporation of genomic DNA spanning a gene of interest as part of the *de novo* AC assembly process. One of the first indications that proximity to centromeric (and telomeric) chromatin may not necessarily be a constraint to vector development came from Bayne and colleagues, who were able to show that expression of a hygromycin resistance gene driven off a strong viral promoter (SV40 early promoter) placed between a functional telomere and centromeric α-satellite sequence on a human X chromosome derivative, was not detectably repressed *(55)*. Selectable marker genes driven by viral promoters are routinely incorporated into chromosome derivatives during telomere-associated chromosome fragmentation. Others have reported retrofitting existing minichromosomes with genes other than selectable markers; for example a 5.5 Mb human chromosome 1-derived minichromosome has been retrofitted with a cDNA version the interleukin-2 gene placed under a viral promoter *(52)*. After transfer of the retrofitted minichromosome into a murine lymphoblastoid cell line by cell fusion sufficient cytokine was produced to relieve the cells of their interleukin-2 dependence, an encouraging demonstration of the feasibility of using engineered minichromosomes as gene vectors.

It has been also been demonstrated that human housekeeping genes can be expressed under their own promoters from relatively small *de novo* ACs. A 404 kb BAC construct, assembled to include a 162 kb genomic fragment containing the hypoxanthine guanine phosphoribosyltransferase (HPRT) gene with its own promoter, 220 kb of human 17 α-satellite DNA and telomere sequences, has been used to generate *de novo* minichromosomes capable of complementing HPRT-deficient HT1080 cells after direct selection in HAT medium *(56)*.

In parallel with this, another group cotransfected a PAC-based vector containing 70 kb of human 21 α-satellite DNA with a second PAC containing a 140 kb genomic fragment encompassing the human HPRT gene. In this experiment clones were selected for based on incorporation of a positive selectable marker and *de novo* ACs composed of both input DNAs were subsequently demonstrated to show persistent transgene expression and to complement the HPRT deficiency in recipient HT1080 cells *(31)*. These reports presented evidence for differences in the structure and mitotic stability of *de novo* ACs that had incorporated the HPRT transgene compared with those generated with alphoid DNA alone: the AC-transgene vectors contained multiple copies of the alphoid input DNA interspersed with large tracts of vector and HPRT sequences and were less mitotically stable. Nevertheless, despite these differences, genes incorporated into small *de novo* ACs are capable of sufficient expression to correct a genetic deficiency in cultured immortal cells.

4. ACs and Minichromosomes as Research Tools

4.1. Chromosome Structure and Function

Minichromosome formation should offer considerable insight into the role of primary DNA sequence and epigenetic factors in centromere formation and maintenance. The assembly of ACs using a variety of different DNA sequences provides a method for the systematic assessment of the relationship between primary DNA sequence and centromere formation. As yet attempts to generate *de novo* centromeres using DNA underlying analphoid neocentromeres have failed to recover ACs *(47)*. However, analphoid centromeres are rare and might be initiated via an uncommon pathway for centromere formation. Clearly the DNA requirements for *de novo* centromere formation are distinct from those involved in the continued maintenance of an existing centromere.

Minichromosomes also offer the potential to duplicate and vary dicentric chromosomal rearrangements to determine how epigenetic modifications are coordinated. In vitro assembly of ACs followed by modifications to replicate the putative epigenetic mechanisms prior to their transfer into cells could also provide a method for actually testing the various proposed epigenetic determinants.

Defined minichromosomes will help answer basic questions about the effects on mitotic stability of decreasing chromosome size and the influence on this of chromosome structure (circular or linear) and centromeric DNA make-up. They may also help to answer other complex questions pertaining to cell cycle regulation, initiation of DNA replication and telomere function. The ability to incorporate DNA fragments of almost unlimited size will make them invaluable tools for examining coordinated regulation of genes clusters, position effects and long range gene regulation, X inactivation, and genomic imprinting.

4.2. Transchromosomal Mice

The generation of transgenic animals carrying extra chromosomes is an exciting prospect allowing for basic research into the chromosomal requirements of meiosis, the creation of mouse models of human genetic diseases and the expression of large human transgenes in animals. Transchromosomal mice have been generated both via the embryonic stem (ES) cell route and through pronuclear injection.

One of the first demonstrations that a single, additional, and structurally defined minichromosome could be passed through the mouse germline was reported by Shen and colleagues *(45)*. A human Y-derived minichromosome was found to be unstable in mouse cells but was rearranged and stabilized through the acquisition of mouse centromeric DNA. This 4.5 Mb linear minichromosome was transferred into mice via the ES cell route, where it was stable and was occasionally transmitted through the germline. Others have introduced modified human marker chromosomes and chromosome fragments into mice via ES cells and have also obtained germline transmission *(57–60)*. A murine satellite DNA artificial chromosome (designated a SATAC) has been successfully transferred into mice by pronuclear injection of flow-sorted chromosomes and germline transmission was obtained *(54)*. It is clear from these reports that the retention of minichromosomes and chromosome fragments in cultured and in somatic cells and the efficiency of germline transmission in transchromosomal mice varies, but as yet the requirements for mitotic and meiotic stability have not been systematically examined. Crucially these studies demonstrate that the presence of an extra univalent chromosome is not necessarily incompatible with meiosis.

Two groups have taken the transgenic approach a step further and have used site-specific recombination (Cre/*loxP*) to retrofit modified human chromosomes with genes of interest: human HPRT *(60)* and the human immunoglobulin (Ig) heavy and λ light chain genes *(61)*. The genes were stably expressed after transfer into mice via ES cells and cell fusion. The latter study, in which human antibodies were produced in mice, illustrates the feasibility of engineering chromosomes to carry megabase-sized inserts. This technology has now been combined with somatic cell nuclear transfer to generate cloned transchromosomal calves producing human immunoglobulin *(62)*.

5. Chromosome-Based Vectors for Human Somatic Gene Therapy?

The correction of human disease by altering the genetic complement of cells is the Holy Grail of modern medicine. For much of the last decade, gene therapy cures have seemed just around the corner. However, technical difficulties in transferring DNA to target tissues and maintaining appropriate expression patterns have been major obstacles. One exciting possibility for gene therapy is the use of chromosome-based gene transfer vectors, which will allow DNA fragments of unlimited to size to be stably maintained in human cells on independently segregating chromosomal elements.

5.1. Ideal Chromosome Vector for Gene Therapy

Any chromosome-based gene transfer vector should have a similar level of structural and mitotic stability to the endogenous human chromosomes. Several studies have observed good structural stability and high retention rates for ACs and minichromosomes; however, some caution is required because these observations were made in cultured immortal cell lines, not in primary somatic cells *(13,14,19,44,46,63)*. Flexibility in the range of DNA fragment sizes that can be incorporated into the AC would be very desirable. This would allow anything from small genes to whole chromosomal regions to be inserted into the chromosome vector. Conceptually and experimentally there does not appear to be any upper limit on the size of the genetic material that can be included, allowing regulatory sequences many tens (or even hundreds) of kilobases away from the body of the gene to be included, thereby maximizing the likelihood of correct temporospatial gene expression. In addition the transfer of entire clusters of genes would provide the potential to correct contiguous gene-deletion syndromes. As yet, whether there is a lower size limit for a stable human chromosome and whether this varies depending on chromosome structure has not been determined. Any chromosome-based vector should be devoid of virally encoded antigens or foreign proteins of any kind that would stimulate the immune system. Most problematic is the desire for any chromosome-based vector to be amenable to easy delivery to the target cell population. The chromosome-based vector itself will have a passive role in the uptake process, carried in by virally or chemically based agents. With current strategies of chromosome engineering and low efficiency delivery techniques, ex vivo gene therapy (where the chromosome vector is both constructed and inserted into patients' cells outside the body followed by reintroduction of the modified cells into the patient) is the only realistic option.

6. Prospects

Our accelerating understanding of chromosome function is rapidly removing the obstacles to efficient generation of artificial chromosomes, with substantial progress made in defining an optimal structure. Although major breakthroughs in manufacture and delivery are essential before application in the gene therapy arena can be realized, nevertheless, this progress is helping to embody artificial chromosomes with many desirable features for basic research in cultured cells and for creating transgenic animals. This an exciting time of discovery, not only for chromosome engineering, but also for developing artificial chromosomes as tools for biology and medicine.

References

1. Murray, A.W. and Szostak, J.W. (1983) Construction of artificial chromosomes in yeast. *Nature* **305**, 189–193.
2. Kipling, D. (ed.) (1995) *The Telomere*. OUP, Oxford.
3. Farr, C., Fantes, J., Goodfellow, P., and Cooke, H. (1991) Functional reintroduction of human telomeres into mammalian cells. *Proc. Natl. Acad. Sci. USA* **88,** 7006–7010.
4. Barnett, M. A., Buckle, V. J., Evans, E. P., Porter, A. C. G., Rout, D., Smith, A. G., et al. (1993) Telomere directed fragmentation of mammalian chromosomes. *Nucleic Acids Res.* **21,** 27–36.
5. Hanish, J.P ., Yanowitz, J. L., and de Lange, T. (1994) Stringent sequence requirements for the formation of human telomeres. *Proc. Natl. Acad. Sci. USA* **91**, 8861–8865.
6. Gilbert, D. M. (2001) Making sense of eukaryotic DNA replication origins. *Science* **294,** 96–100.
7. Choo, K. H. A. (ed.) (1997) *The Centromere*. OUP, Oxford.

8. Oakey, R. and Tyler-Smith, C. (1990) Y chromosome DNA haplotyping suggests that most European and Asian men are descended from one of two males. *Genomics* **7,** 325–330.

9. Mahtani, M. M. and Willard, H. F. (1998) Physical and genetic mapping of the human X chromosome centromere: repression of recombination. *Genome Res.* **8,** 100–110.

10. Lo, A. W. I., Liao, G. C. C., Rocchi, M., and Choo, K. H. A. (1999) Extreme reduction of chromosome-specific α-satellite array is unusually common in human chromosome 21. *Genome Res.* **9,** 895–908.

11. Choo, K. H. (2000) Centromerization. *Trends Cell Biol.* **10,** 182–188.

12. Warburton, P. E. (2001) Epigenetic analysis of kinetochore assembly on variant human centromeres. *Trends Genet.* **17,** 243–247.

13. Harrington, J. J., Van Bokkelen, G., Mays, R. W., Gustashaw, K., and Willard, H. F. (1997) Formation of de novo centromeres and construction of first-generation human artificial microchromosomes. *Nat. Genet.* **15,** 345–355.

14. Ikeno, M., Grimes, B., Okazaki, T., Nakano, M., Saitoh, K., Hoshino, H., et al. (1998) Construction of YAC-based mammalian artificial chromosomes. *Nat. Biotech.* **16,** 431–439.

15. Henning, K., Novotny, E., Compton, S., Guan, X., Liu, P., and Ashlock, M. (1999) Human artificial chromosomes generated by modification of a yeast artificial chromosome containing both human alpha satellite and single-copy DNA sequences. *Proc. Natl. Acad. Sci. USA* **96,** 592–597.

16. Masumoto, H., Ikeno, M., Nakano, M., Okazaki, T., Grimes, B., Cooke, H., and Suzuki, N. (1999) Assay of centromere function using a human artificial chromosome. *Chromosoma* **107,** 406–416.

17. Ebersole, T., Ross, A., Clark, E., McGill, N., Schindelhauer, D., Cooke, H., and Grimes, B. (2000) Mammalian artificial chromosome formation from circular alphoid input DNA does not require telomere repeats. *Hum. Mol. Genet.* **9,** 1623–1631.

18. Schueler, M. G., Higgins, A. W., Rudd, M. K., Gustashaw, K., and Willard, H. F. (2001) Genomic and genetic definition of a functional human centromere. *Science* **294,** 109–115.

19. Mejia, J. E., Alazami, A., Willmott, A., Marschall, P., Levy, E., Earnshaw, W. C., and Larin, Z. (2002) Efficiency of de novo centromere formation in human artificial chromosomes. *Genomics* **79,** 297–304.

                                                                                                  Lim and Farr

20. Ohzeki, J.-I., Nakano, M., Okada, T., and Masumoto, H. (2002) CENP-B box is required for de novo centromere chromatin assembly on human alphoid DNA. *J Cell Biol.* **159,** 765–775.
21. Kouprina, N., Ebersole, T., Koriabine, M., Pak, E., Rogozin, I. B., Katoh, M., et al. (2003) Cloning of human centromeres by transformation-associated recombination in yeast and generation of functional human artificial chromosomes. *Nucleic Acids Res.* **31,** 922–934.
22. Hudson, D. F., Fowler, K. J., Earle, E., Saffery, R., Kalitsis, P., Trowell, H., et al. (1998) Centromere protein B null mice are mitotically and meiotically normal but have lower body and testis weights. *J. Cell Biol.* **141,** 309–319.
23. Kapoor, M., Montes de Oca Luna, R., Liu, G., Lozano, G., Cummings, C., Mancini, M., et al. (1998) The cenpB gene is not essential in mice. *Chromosoma.* **107,** 570–576.
24. Perez-Castro, A. V., Shamanski, F. L., Meneses, J. J., Lovato, T. L., Vogel, K. G., Moyzis, R. K., and Pedersen, R. (1998) Centromeric protein B null mice are viable with no apparent abnormalities. *Dev. Biol.* **32,** 135–143.
25. Tomascik-Cheeseman, L., Marchetti, F., Lowe, X., Shamanski, F. L., Nath, J., Pedersen, R. A., and Wyrobek, A. J. (2002) CENP-B is not critical for meiotic chromosome segregation in male mice. *Mutat. Res.* **513,** 197–203.
26. Goldberg, I. G., Sawhney, H., Pluta, A. F., Warburton, P. E., and Earnshaw, W. C. (1996) Surprising deficiency of CENP-B binding sites in African green monkey alpha-satellite DNA: implications for CENP-B function at centromeres. *Mol. Cell Biol.* **16,** 5156–5168.
27. Yoda, K., Nakamura, T., Masumoto, H., Suzuki, N., Kitagawa, K., Nakano, M., et al. (1996) Centromere protein B of African green monkey cells: gene structure, cellular expression, and centromeric localization. *Mol. Cell Biol.* **16,** 5169–5177.
28. Barry, A. E., Bateman, M. A., Howman, E. V., Cancilla, M. R., Tainton, K. M., Irvine, D. V., et al. (2000) The 10q25 neocentromere and its inactive progenitor have identical primary nucleotide sequence: Further evidence for epigenetic modification. *Genome Res.* **10,** 832–838.
29. Saffery, R., Irvine, D. V., Griffiths, B., Kalitsis, P., Wordeman, L., and Choo, K. H. (2000) Human centromeres and neocentromeres show identical distribution patterns of >20 functionally important kinetochore-associated proteins. *Hum. Mol. Genet.* **9,** 175–185.

30. Ando, S., Yang, H., Nozaki, N., Okazaki, T., and Yoda, K. (2002) CENP-A, -B, and -C chromatin complex that contains the I-type alpha-satellite array constitutes the prekinetochore in HeLa cells. *Mol. Cell Biol.* **22,** 2229–2241.

31. Grimes, B. R., Schindelhauer, D., McGill, N. I., Ross, A., Ebersole, T. A., and Cooke, H. J. (2001) Stable gene expression from a mammalian artificial chromosome. *EMBO Rep.* **2,** 910–914.

32. Farr, C. J. Chromosome fragmentation in vertebrate cell lines, in *Chromosome Structural Analysis,* (Bickmore, W. A., ed.), OUP, Oxford, 1999, pp. 183–198.

33. Farr, C. J., Stevanovic, M., Thomson, E. J., Goodfellow, P. N., and Cooke, H. J. (1992) Telomere-associated chromosome fragmentation: applications in genome manipulation and analysis. *Nat. Genet.* **2,** 275–282.

34. Au, H. C., Mascarello, J. T., and Scheffler, I. E. (1999) Targeted integration of a dominant neo(R) marker into a 2 to 3 Mb human minichromosome and transfer between cells. *Cytogenet. Cell Genet.* **86,** 194–203.

35. Hernandez, D., Mee, P. J., Martin, J. E., Tybulewicz, V. L., and Fisher, E. M. (1999) Transchromosomal mouse embryonic stem cell lines and chimeric mice that contain freely segregating segments of human chromosome 21. *Hum. Mol. Genet.* **8,** 923–933.

36. Moralli, D., Vagnarelli, P., Bensi, M., De Carli, L., and Raimondi, E. (2001) Insertion of a loxP site in a size-reduced human accessory chromosome. *Cytogenet. Cell Genet.* **94,** 113–120.

37. Buerstedde, J.-M. and Takeda, S. (1991) Increased ratio of targeted to random integration after transfection of chicken B cell lines. *Cell* **67,** 179–188.

38. Dieken, E. S., Epner, E. M., Fiering, S., Fournier, R. E. K., and Groudine, M. (1996) Efficient modification of human chromosomal alleles using recombination-proficient chicken/human microcell hybrids. *Nat. Genet.* **12,** 174–182.

39. Dieken, E. S. and Fournier, R. E. K. (1996) Homologous modification of human *Chromosomal* genes in chicken B-cell x human microcell hybrids, in *Methods* (Fournier, R. E.K., ed.), AP, San Diego, 1996, pp. 56–63.

40. Koi, M., Lamb, P. W., Filatov, L., Feinberg, A. P., and Barrett, J. C. (1997) Construction of chicken-human microcell hybrids for human gene targeting. *Cytogenet. Cell Genet.* **76,** 72–76.

41. Kuroiwa, Y., Shinohara, T., Notsu, T., Tomizuka, K., Yoshida, H., Takeda, S.-i., et al. (1998) Efficient modification of a human chromosome by telomere-directed truncation in high homologous recombination-proficient chicken DT40 cells. *Nucleic Acids Res.* **26,** 3447–3448.

42. Shen, M. H., Yang, Y., Loupart, M.-L., Smith, A., and Brown, W. (1997) Human mini-chromosomes in mouse embryonal stem cells. *Hum. Mol. Genet.* **6,** 1375–1382.

43. Farr, C. J., Bayne, R. A. L., Kipling, D., Mills, W., Critcher, R., and Cooke, H. J. (1995) Generation of a human X-derived minichromosome using telomere-associated chromosome fragmentation. *EMBO J.* **14,** 5444–5454.

44. Mills, W., Critcher, R., Lee, C., and Farr, C. (1999) Generation of an ~2.4 Mb human X centromere-based minichromosome by targeted telomere-associated chromosome fragmentation. *Hum Mol Genet.* **8,** 751–761.

45. Shen, M. H., Mee, P. J., Nichols, J., Yang, J., Brook, F., Gardner, R. L., et al. (2000) A structurally defined mini-chromosome vector for the mouse germ line. *Curr. Biol.* **10,** 31–34.

46. Yang, J., Pendon, C., Yang, J., Haywood, N., Chand, A., and Brown, W. R. (2000) Human mini-chromosomes with minimal centromeres. *Hum. Mol. Genet.* **9,** 1891–1902.

47. Saffery, R., Wong, L. H., Irvine, D. V., Bateman, M. A., Griffiths, B., Cutts, S. M., et al. (2001) Construction of neocentromere-based human minichromosomes by telomere-associated chromosomal truncation. *Proc. Natl. Acad. Sci. USA* **98,** 5705–5710.

48. Spence, J. M., Critcher, R., Ebersole, T. A., Valdivia, M. M., Earnshaw, W. C., Fukagawa, T., Farr, C. J. (200) Co-localization of centromere activity, proteins and topoismerase II within a subdomain of the major human X alpha-satellite array. *EMBO J.* **21,** 5269–5280.

49. Kereso, J., Praznovszky, T., Cserpan, I., Fodor, K., Katona, R., Csonka, E., et al. (1996) De novo chromosome formation by large scale amplification of the centromeric region of mouse chromosomes. *Chrom. Res.* **4,** 226–239.

50. Csonka, E., Cserpan, I., Fodor, K., Hollo, G., Katona, R., Kereso, J., et al. (2000) Novel generation of human satellite DNA-based artificial chromosomes. *J. Cell Sci.* **113,** 3207–3216.

51. Raimondi, E., Balzaretti, M., Moralli, D., Vagnarelli, P., Tredici, F., Bensi, M., et al. (1996) Gene targeting to the centromeric DNA of a human minichromosome. *Hum. Gene Ther.* **7,** 1103–1109.

52. Guiducci, C., Ascenzioni, F., Auriche, C., Piccolella, E., Guerrini, A., and Donini, P. (1999) Use of a human minichromosome as a cloning and expression vector for mammalian cells. *Hum. Mol. Genet.* **8,** 1417–1424.

53. Vanderbyl, S., MacDonald, N., and De Jong, G. (2001) A flow cytometry technique for measuring chromosome-mediated gene transfer. *Cytometry* **44,** 100–105.

54. Co, D., Borowski, A., Leung, J., van der Kaa, J., Hengst, S., Platenburg, G., et al. (2000) Generation of transgenic mice and germline transmission of a mammalian artificial chromosome introduced into embryos by pronuclear microinjection. *Chrom. Res.* **8,** 183–191.

55. Bayne, R. A. L., Broccoli, D., Taggart, M. H., Thomson, E. J., Farr, C. J., and Cooke, H. J. (1994) Sandwiching of a gene within 12 kb of a functional telomere and alpha satellite does not result in silencing. *Hum. Mol. Genet.* **3,** 539–546.

56. Mejia, J. E., Willmott, A., Levy, E., Earnshaw, W. C., and Larin, Z. (2001) HAC-mediated rescue of HPRT deficiency. *Am. J. Hum. Genet.* **69,** 315–326.

57. Tomizuka, K., Yoshida, H., Uejima, H., Kugoh, H., K., S., Ohguma, A., Hayasaka, M., et al. (1997) Functional expression and germline transmission of a human chromsome fragment in chimaeric mice. *Nat. Genet.* **16,** 133–143.

58. Kazuki, Y., Shinohara, T., Tomizuka, K., Katoh, M., Ohguma, A., Ishida, I., et al. (2001) Germline transmission of a transferred human chromosome 21 fragment in transchromosomal mice. *J. Hum. Genet.* **46,** 600–603.

59. Shinohara, T., Tomizuka, K., Miyabara, S., Takehara, S., Kazuki, Y., Inoue, J., et al. (2001) Mice containing a human chromosome 21 model behavioral impairment and cardiac anomalies of Down's syndrome. *Hum. Mol. Genet.* **10,** 1163–1175.

60. Voet, T., Vermeesch, J., Carens, A., Durr, J., Labaere, C., Duhamel, H., et al. (2001) Efficient male and female germline transmission of a human chromosomal vector in mice. *Genome Res.* **11,** 124–136.

61. Kuroiwa, Y., Tomizuka, K., Shinohara, T., Kazuki, Y., Yoshida, H., Ohguma, A., et al. (2000) Manipulation of human minichromosomes to carry greater than megabase-sized chromosome inserts. *Nat. Biotech.* **18,** 1086–1090.

62. Kuroiwa, Y., Kasinathan, P., Choi, Y. J., Naeem, R., Tomizuka, K., Sullivan, E. J., et al. (2002) Cloned transchromosomic calves producing human immunoglobulin. *Nat. Biotech.* **20,** 889–894.

63. Shen, M., Yang, J., Pendon, C., and Brown, W. (2001) The accuracy of segregation of human mini-chrommosomes varies in different vertebrate cell lines, correlates with the extent of centromere formation and provides evidence for a trans-acting centromere maintenance activity. *Chromosoma* **109,** 524–535.

10

Mammalian Artificial Chromosome Formation in Human Cells After Lipofection of a PAC Precursor

Jose I. de las Heras, Leonardo D'Aiuto, and Howard Cooke

1. Introduction

An artificial chromosome is a synthetic structure that carries three fundamental components for its long-term survival, replication, and segregation after cell division. These components are telomeres, one or more replication origins, and a centromere. The creation of such a molecule became feasible initially in the budding yeast *Saccharomyces cerevisiae*, where replication origins and centromeric sequences are well defined. In this organism, autonomously replicating sequences (ARS) were isolated by their ability to allow replication of plasmids carrying them. Plasmids containing an ARS element are capable of extrachromosomal replication in selective conditions but are lost if selection is removed from the culture because of unequal segregation. Introduction of a functional centromere to an ARS plasmid provides mitotic stability to the resulting yeast artificial chromosome (YAC) (reviewed by Newlon in **ref. 1**).

In contrast with *S. cerevisiae*, the DNA sequences necessary for replication and segregation in mammalian cells are poorly under-

From: *Methods in Molecular Biology, Vol. 240:*
Mammalian Artificial Chromosomes: Methods and Protocols
Edited by: V. Sgaramella and S. Eridani © Humana Press Inc., Totowa, NJ

stood. In particular, the nature and indeed the existence of the *cis*-acting DNA sequence necessary for centromerization in mammalian cells remains controversial. To fill these gaps, much work has been dedicated to developing strategies for the construction of mammalian artificial chromosomes (MACs). One reason for these efforts is that MACs could be very useful tools to elucidate basic aspects of mammalian chromosome function and structure. In addition, the ability to produce MACs would have important implications in the field of gene therapy. Effective therapy by in vivo delivery of DNA requires efficient delivery, long-term maintenance, and expression of the therapeutic gene in the target cells. Full controlled expression at physiological levels can be achieved after the transfer of intact genes, together with their regulatory regions, spanning hundreds of kilobases. Long-term maintenance and segregation in mammalian cells of the input DNA could be attained if the DNA carried replication origins, a centromere, and telomeres. These features may be combined in a MAC or minichromosome, which would make them ideal vectors for gene therapy.

Two approaches to the construction of the above chromosomal structures are being pursued.

In the top-down approach, cloned telomeric DNA is used to fragment a natural chromosome into a minichromosome through one or more rounds of telomere-directed chromosome breakage. This strategy has been applied successfully to generate: (1) a series of linear minichromosomes based on the human X centromere *(2)*; (2) a 5.5-mb minichromosome derived from human chromosome 1 *(3)*; and (3) a 4.5-Mb minichromosome named St1, which originated when a human Y chromosome truncated derivative rearranged in a mouse cell line, acquiring mouse centromeric satellite DNA in the process *(4,5)*. Minichromosomes have also been created in some lower organisms, including the fruit fly *Drosophila melanogaster (6)* and the parasitic protozoan *Trypanosoma brucei (7)*.

The bottom-up approach is based on the assembly of MACs from cloned elements combined in the laboratory and then introduced into mammalian cells. Harrington et al. *(8)* synthesized α-satellite arrays through the multimerization of a single α-satellite higher order repeat unit from human chromosomes 17 and Y. After introducing these arrays into the human cell line HT1080, combined with telomeric and fragmented human genomic DNA, they obtained artificial chromosomes with an estimated size of 6–10 Mb. Following a similar strategy, different α-satellite arrays have been tested for their ability to induce *de novo* chromosome formation. Ikeno et al. used two 100-kb YACs containing two different subfamilies of α-satellite repeats from human chromosome 21 propagated in a recombination deficient host strain (rad52) *(9)*. To overcome instability of YACs containing heterochromatic sequences, telomeres and selectable markers were retrofitted in these clones after transiently expressing the protein Rad52 required for recombination. Only one of the two YACs (α7c5hTEL) was able to form MACs in HT1080 *(9)*. In another example, a MAC was produced by transfection of a YAC containing satellite sequences (including α-satellite) and modified to contain human telomeric DNA and a putative origin of replication from the human β-globin locus *(10)*. Recently, our laboratory has been able to efficiently produce MACs from PAC (phage P1-based artificial chromosome) vectors containing α-satellite derived from the YAC construct α7c5hTEL *(11)*. Upon lipofection of the PAC construct pTAT/7c5/BS (*see* **Fig. 1**) in the cell line HT1080, *de novo* artificial chromosomes were observed in at least 50% of the resulting clones (*see* **Fig. 2**). Of these, many contained a MAC in 80% or more of the cells *(11)*.

In this chapter, we describe in detail the procedure to induce MAC formation in HT1080 cell line by means of the lipofection of the pTAT/7c5/BS construct. These methods are applicable to any circular construct grown in a bacterial host.

2. Materials

2.1. PAC DNA Extraction

This method follows a modified protocol from the standard Qiagen Maxiprep kit, where most of the materials listed can be obtained (*see* **Note 1**).

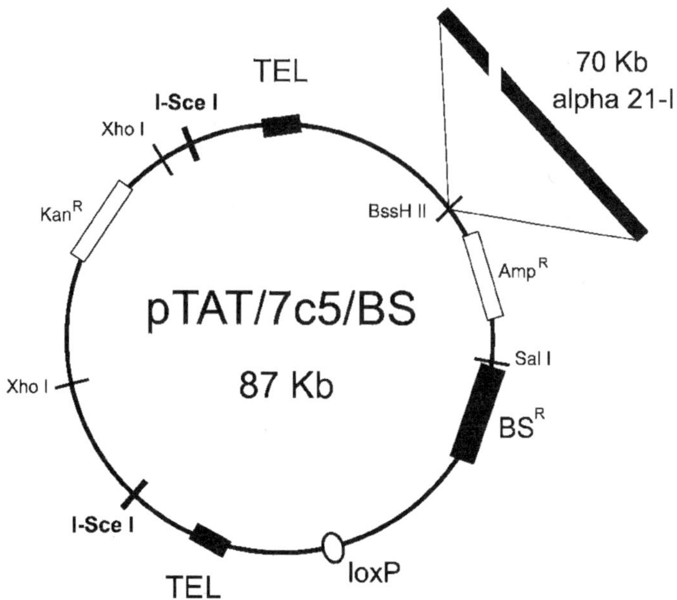

Fig. 1. Diagram of the PAC construct pTAT/7c5/BS, which has the ability to induce *de novo* formation of MACs in HT1080 cells. Its main components are (1) two 800-bp telomeric cassettes inserted in opposite orientations (TEL); (2) blasticidin resistance gene (BS^R), for selection in mammalian cells; and (3) 70 kb of centromeric α-satellite DNA from human chromosome 21 (α21-I). This satellite DNA is also thought to provide the necessary replication origins in human cells *(12)*. Other components include a loxP site for the in vivo insertion of genes using the Cre-Lox system and the unique restriction enzyme site *Sal*I, which can be used for the cloning of additional elements. This construct can be linearized by digestion with either *Xho*I or the very rare cutting enzyme I-*Sce*I, leaving a linear MAC precursor with telomeric cassettes at both ends. Both linear forms and circular forms are able to induce *de novo* MAC formation *(11)*. The PAC backbone is based on pPAC4 *(13)*.

1. Sorvall rotors SLA-1500 and SS-34 or equivalents.
2. 250-mL centrifuge bottles to fit rotor SLA-1500, autoclaved.
3. 30-mL Corex tubes and rubber adaptors to fit rotor SS-34, baked.
4. Plastic funnels.
5. Whatman No. 1 filter paper.
6. 50-mL Falcon tubes.

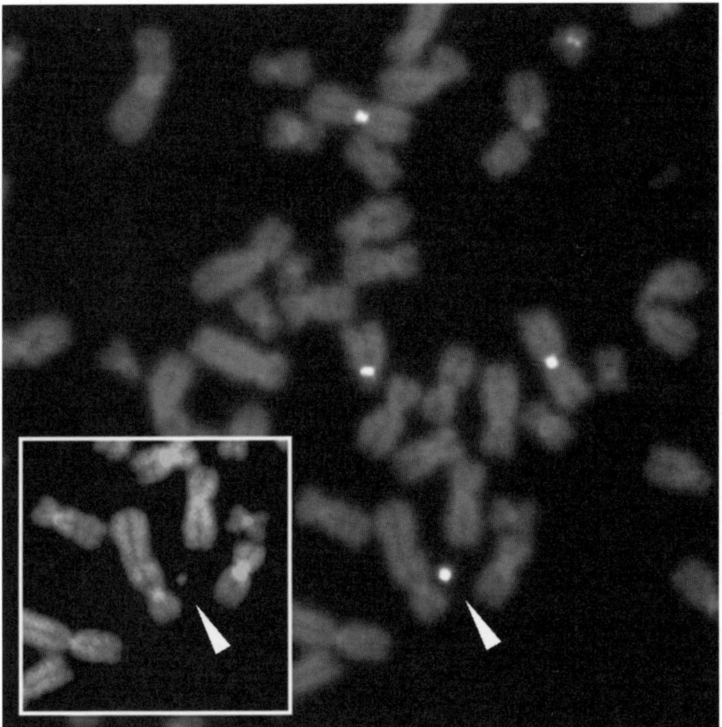

Fig. 2. Photograph of a HT1080 metaphase after FISH using the probe p11.4 specific for α-satellite DNA (*see* **Note 18**). The main picture shows the p11.4 probe in white with the chromosomes counterstained with DAPI, whereas the inset shows only the DAPI signal. A small MAC is indicated by an arrowhead.

7. Box of autoclaved 200-μL pipet tips with the ends cut (wide bore).
8. Solution P1 (resuspension buffer): 50 m*M* Tris-HCl, pH 8.0; 10 m*M* ethylenediamine tetraacetic acid.
9. RNAse A, 100 mg/mL stock (Sigma).
10. Solution P2 (lysis buffer): 200 m*M* NaOH; 1% sodium dodecyl sulfate (SDS).
11. Solution P3 (neutralization buffer): 3 *M* potassium acetate, pH 5.5.
12. Q500 resin column (Qiagen GmbH, Germany).
13. 2% SDS solution.
14. Solution QBT (equilibration buffer): 750 m*M* NaCl; 50 m*M* MOPS, pH 7.0; 15% isopropanol; 0.15% Triton X-100.

15. Solution QC (wash buffer): 1 *M* NaCl; 50 m*M* MOPS, pH 7.0; 15% isopropanol.
16. Solution QF (elution buffer): 1.25 *M* NaCl; 50 m*M* Tris-HCl, pH 8.5; 15% isopropanol.
17. Isopropanol (ice cold).
18. 70% Ethanol.

2.2. PAC DNA Purification for Lipofection

1. Pulsed field gel electrophoresis apparatus.
2. Scalpel and small spatula.
3. 2-mL microfuge tubes.
4. 45°C and 65°C waterbaths.
5. Low melting point agarose (Ultrapure LMP Agarose. GibcoBRL).
6. 0.25 TBE buffer (dilute from 20 stock: 1.78 *M* Tris-HCl pH 8.3; 1.78 *M* boric acid; 100 m*M* ethylenediamine tetraacetic acid).
7. Ethidium bromide 10 mg/mL stock.
8. Agarase (1 U/µL) and 25 agarase buffer (Roche, Germany).
9. Phenol:chloroform:isoamyl alcohol (25:24:1; Sigma).
10. Isopropanol.
11. 70% Ethanol.

2.3. Tissue Culture and Lipofection

Standard tissue culture tools and materials are not listed.

1. 25-cm^2 and 75-cm^2 tissue culture flasks.
2. HT1080 human fibrosarcoma cells in culture.
3. Dulbecco's modified Eagle medium culture medium (GibcoBRL) supplemented with 10% bovine calf serum (Hyclone, UT), and penicillin/streptomycin (100 U/mL penicillin; 100 µg/mL streptomycin).
4. Optimem I (GibcoBRL; serum-free culture medium).
5. Phosphate buffered saline solution (PBS; 150 m*M* NaCl; 3 m*M* Na_2HPO_4; 1 m*M* KH_2PO_4).
6. Blasticidin S Hydrochloride (ICN Biochemicals, OH). Stock 100 mg/mL stored at 4°C.
7. Lipofectamine Plus reagents (GibcoBRL).

2.4. FISH Analysis

2.4.1. Chromosome Preparation

1. Phase-contrast microscope.
2. Waterbath at 37°C.
3. 15-mL Falcon tubes.
4. Benchtop centrifuge with adaptors for 15-mL Falcon tubes.
5. Disposable 1-mL plastic pipets.
6. Colcemid (marketed as demecolcine; Sigma) 10 µg/mL stock.
7. Warm (37°C) hypotonic solution: 75 mM KCl.
8. Ice-cold fixative 3:1 methanol:glacial acetic acid. Freshly prepared.

2.4.2. Slide Preparation

1. Glass microscope slides. Kept stored in ethanol.
2. Vacuum chamber with slide rack. Vacuum pump.
3. Incubator at 37°C to store the vacuum chamber.

2.4.3. Probe Labeling for Fluorescence In Situ *Hybridization (FISH)*

1. Waterbath at 16°C.
2. Nick translation (NT) buffer, 10X stock (0.5 M Tris-Cl, pH 7.8; 50 mM MgCl$_2$; 10 mM β-mercaptoethanol; 500 µg/mL bovine serum albumin).
3. AGC nucleotide mix (dATP, dGTP, dCTP 0.5 mM stock; ABgene, UK).
4. dTTP (0.5 mM stock; ABgene, UK).
5. Biotin-16-dUTP 1 mM (Roche, Germany).
6. Digoxigenin -11-dUTP 1 mM (Roche, Germany).
7. DNAse I 10 U/µL (Roche, Germany).
8. Polymerase I (10 U/µL; GibcoBRL).
9. DNA for vector probe: pPAC4 (PAC vector).
10. DNA for α13/21: p11.4 (specific at high stringency for alphoid arrays present in chromosomes 13 and 21, *see* **Note 18**).

                                                          de las Heras et al.

2.4.4. Hybridization

1. Waterbaths at 37 and 73°C.
2. Staining trough with 2X SSC (20X SSC: 0.3 M sodium citrate; 3 M NaCl; pH 7.0) plus 25 μg/mL RNAseA at 37°C.
3. Staining troughs with 70, 90, and 100% ethanol.
4. Empty staining trough in the 73°C waterbath.
5. Metal tray in a 37°C waterbath covered with aluminium foil.
6. Vacuum chamber with slide rack.
7. Speedvac machine.
8. Coverslips, 22 × 22 mm kept stored in a 50-mL Falcon tube with ethanol.
9. Rubber solution to seal coverslips.
10. Sonicated salmon sperm (10 mg/mL).
11. Hybridization solution (50% deionized formamide; 20% dextran sulphate; 2X SSC; 1% Tween 20).
12. 70% Formamide/4X SSC, freshly prepared, prewarmed at 73°C for chromosome denaturation (*see* **Note 2**).

2.4.5. Stringency Washes and Label Detection

1. Waterbaths at 37, 45, and 60°C, with an empty staining trough in each.
2. Coverslips, 22 × 40 mm. Kept stored in a 50-mL Falcon tube with ethanol.
3. 2X SSC at 45°C.
4. 0.5X SSC at 60°C.
5. 4X SSC/0.1% Tween 20 at 37°C.
6. 4X SSC.
7. Blocking reagent (5% skimmed milk in 4X SSC).
8. Rhodamine conjugated anti-digoxigenin (sheep Fab fragments) 200 μg/mL stock (Roche, Germany).
9. Fluorescein avidin DCS (cell sorter grade) 2 mg/mL stock (Vector Laboratories).
10. DAPI/Vectashield mountant. Make a 50 μg/mL stock of DAPI (Sigma) in distilled water. Dilute stock 1/25 with Vectashield (Vector Laboratories) and store at 4°C (*see* **Note 3**).

3. Methods

3.1. PAC DNA Extraction

The construct pTAT/7c5/BS (*see* **Fig. 1**) was maintained in the *Escherichia coli* strain DH10β. DNA was extracted using the Maxiprep kit from Qiagen, according to the following protocol, modified from the manufacturer's instructions (*see* **Note 4**).

1. Distribute 800–1200 mL overnight culture in 250-mL centrifuge bottles, to fit Sorvall rotor SLA-1500 or equivalent (*see* **Note 5**).
2. Centrifuge at 4°C for 10 min at 4500 rpm (3000g max.).
3. Discard supernatant and resuspend each pellet in 10 mL solution P1, containing 100 μg/mL RNAse A (*see* **Note 6**).
4. Add 10mL lysis solution P2 to each bottle, mixing well but gently and let stand 5 min at room temperature.
5. Add 10mL ice cold solution P3 and let stand on ice for 30–40 min, mixing ocassionally in a gentle manner.
6. Meanwhile, wash three to six funnels with 2% SDS, rinse well in distilled water, and air-dry.
7. Centrifuge bottles at 4°C and 6000 rpm (5400g max.) for 15 min.
8. Meanwhile, equilibrate a single Q500 Qiagen column with 10 mL solution QBT, placed on a 500-mL conical flask. Also, prepare Whatman No. 1 filters on the funnels, and prewet in solution QBT. (*see* **Note 7**).
9. Filter supernatant through filters into 50-mL Falcon tubes. Pour filtered solution gradually into Q500 column.
10. When all the solution has been loaded into the column, wash twice with 30 mL solution QC.
11. Elute DNA with 15 mL of hot (65°C) solution QF into a 30-mL baked Corex tube.
12. Precipitate DNA with 12 mL cold isopropanol, mixing very gently by inversion. Better yields are obtained after overnight precipitation at 4°C, but 30–60 min is usually enough.
13. Centrifuge tubes at 4°C using Sorvall rotor SS-34 or equivalent, for 30 min at 12,500 rpm (18,700g max.). Make a mark on the tubes to aid locating the pellet after centrifugation.

14. Discard supernatant carefully. Wash with 2 mL of 70% ethanol. Centrifuge as previously for 10 min.

15. Pipet out the supernatant, carefully, first with 1-mL tips and then with 200-µL tips, to remove as much liquid as possible without disturbing the DNA pellet.

16. Air-dry for 5 min and resuspend in 300–400 µL sterile distilled water (*see* **Note 8**). The solution will appear very viscous.

17. Transfer DNA solution to a 1.5-mL microfuge tube using wide bore tips and measure concentration on the spectrophotometer.

18. Optional: washing the walls of the tube again with 100 µL of sterile water will allow recovery of some more DNA, typically at approx 150–200 ng/µL.

3.2. PAC DNA Purification for Lipofection

Circular or linear DNA (i.e., after digestion with the restriction enzymes I-*Sce*I or *Xho*I; *see* **Fig. 1**) can be isolated from a low melting point (LMP) agarose gel. In addition, purification of PAC DNA from a gel results in enhanced MAC formation rates, as compared with DNA purified by other methods (*see* **Note 9**).

3.2.1. Pulsed Field Gel Electrophoresis (PFGE)

1. Cast a 1% LMP agarose gel in 0.25X TBE buffer. Make a large band by taping together several comb teeth as appropriate, enough to hold about 30–40 µg (*see* **Note 10**) of PAC DNA without overloading the gel (*see* **Note 11**).

2. Load samples inside the PFGE apparatus containing 0.25X TBE buffer. Turn buffer pump off to avoid disturbing the wells while loading. A small amount of the PAC DNA solution is also loaded in a normal size well next to the large band containing the bulk of the DNA.

3. Prerun the gel, with the buffer pump off, for 30–40 min at 100 V as in a conventional electrophoresis run (set to constant angle, positive electrode at the bottom of the gel) to allow the samples enter the gel.

4. Turn buffer pump on again, and run PFGE. We use a Rotaphor Type V apparatus (Biometra, Germany), with 30 cm between electrodes. For a construct of the size of pTAT/7c5/BS we use the following conditions.

　　　　Run length: 20 h
　　　　Temperature: 11°C
　　　　Voltage: constant 180 V
　　　　Angle: constant 120°
　　　　Switch times: 10–30 s, linear ramp

5. After the run is complete, cut a vertical strip of agarose containing the large band, and stain the rest of the gel for 30 min in an ethidium bromide bath (20 µL of 10 mg/mL stock solution into 600 mL of electrophoresis buffer) (*see* **Note 12**).
6. Locate position of DNA and cut a mark on the edge of the stained gel. Then, use this mark to position the band on the unstained strip of agarose. Cut agarose-containing DNA of interest with a clean scalpel. Weigh agarose.

3.2.2. DNA Extraction From the LMP Gel

We use Agarase (Roche, Germany) to digest the agarose and recover the DNA.

1. Cut agarose into pieces suitable for treatment in individual tubes. For a 2-mL microfuge tube, pieces of up to 1 mg are adequate.
2. Add 25X agarose buffer to the 2-mL microfuge tubes as required and melt in a 65°C waterbath. Make sure the agarose is completely molten (*see* **Note 13**).
3. Transfer to a 45°C waterbath for 5 min to equilibrate temperature and then add agarase (about 2 U/100 mg of agarose) to the tubes. Mix well and incubate for 90 min (*see* **Note 14**).
4. Check the agarose digestion by incubating the tubes on ice for 5 min. If they do not remain liquid, melt again, and repeat digestion after adding more agarose.
5. Extract DNA with phenol/chloroform, mixing very gently to avoid damage to the high molecular weight DNA. Collect supernatant into a fresh 2-mL microfuge tube.
6. Precipitate DNA with isopropanol at 4°C and centrifuge (*see* **Note 15**). Wash pellet with 70% ethanol, spin again, and resuspend in 20–40 µL sterile distilled water.
7. Measure DNA concentration.
8. Optional: run a small aliquot on a gel to check DNA has not degraded significantly during these manipulations.

3.3. Tissue Culture and Lipofection

1. Split HT1080 cells the day prior to lipofection into 6-well plates, at different densities, aiming to obtain enough wells at approx 50–70% confluency the next day (*see* **Note 16**).
2. Select wells for lipofection. Include a negative control (no input DNA) to check the action of the selection.
3. Prepare pairs of 1.5-mL microfuge tubes for each lipofection. In tube A mix DNA (1–2 µg, *see* **Note 17**) and PLUS reagent (6 µL per µg DNA) with 100 µL Optimem (serum-free medium). In tube B dilute lipofectamine (5 µL/µg of DNA) in 100 µL Optimem. Let stand both tubes for 15 min at room temperature.
4. Mix tube A with B, and let stand another 15 min at room temperature.
5. Remove medium and wash cells with prewarmed (37°C) PBS, then add 800 µL prewarmed Optimem to each well.
6. Add lipofection mixture to each well, mix gently, and incubate cells for 3–5 h.
7. Add 4 mL normal culture medium (complete, without selection) and incubate overnight.
8. Day 1 postlipofection. Remove medium, wash with PBS, and add 3–4 mL normal medium.
9. Day 2 postlipofection. Expand to 25–75-cm^2 tissue culture flasks.
10. Day 3 postlipofection. Apply selection: blasticidine 4 µg/mL.
11. Check the negative control to see when blasticidine kills the cells. Usually clones can be picked up after 2–3 wk after lipofection.
12. Collect clones into 24-well plates and expand into tissue culture flasks. Freeze cells from a confluent 25-cm^2 flask and prepare chromosomes from a 75-cm^2 flask for subsequent FISH analysis.

3.4. FISH Analysis

Chromosome preparations from the clones obtained are routinely analyzed by two-color FISH using probes specific for vector sequences (pPAC4) and for the alphoid array (p11.4) (*see* **Note 18**).

3.4.1. Chromosome Preparation

1. Split cells into a 75-cm^2 flask 1 d in advance. The aim is to obtain an actively dividing cell monolayer at approx 70–80% confluency for chromosome preparation.

2. Add colcemid to the culture medium to a final concentration of 0.1 µg/mL. Incubate at 37°C for 2 h (*see* **Note 19**).
3. Shake flask to dislodge cells, collect medium in a 15-mL benchtop centrifuge tube, and pellet cells at about 200g (*see* **Note 20**) for 4 min.
4. Remove supernatant.
5. Add a few milliliters of PBS to the flask, shake again, and collect into the same 15-mL tube. Repeat centrifugation.
6. Remove supernatant, leaving only enough liquid to resuspend the cells.
7. Add 10 mL prewarmed (37°C) hypotonic solution (75 mM KCl) gently, making sure the cells are well resuspended. Incubate at 37°C for 12 min (*see* **Note 21**).
8. Centrifuge cells and discard supernatant as above. Resuspend gently the cells in the small volume of liquid remaining.
9. Add 10 mL fixative (3:1 methanol:acetic acid, prechilled in ice) dropwise at first, gently mixing with the cells.
10. Centrifuge cells and discard supernatant.
11. Finally, fix the cells by adding another 10 mL ice-cold fixative. Store at 20°C until needed (*see* **Note 22**).

3.4.2. Slide Preparation

For FISH, we usually prepare the slides 1 d in advance.

1. Concentrate cell suspension. If there are enough cells, this is easily done by carefully pipetting out 5–6 mL of the fixative, as cells sediment while in storage at the bottom of the tubes. Otherwise, a brief centrifugation will help minimize cell loss (1 min at 200g is sufficient), and then remove enough supernatant to leave a concentrated suspension. Keep tubes in ice (*see* **Note 23**).
2. Wipe slides with ethanol by rubbing energically with a lint-free cloth. Label slides with a soft-lead pencil.
3. Let a drop of cold cell suspension fall onto the slide from a height of approx 10–15 cm. Wash three or four times the slide with fixative over a beaker using a disposable plastic pipet (*see* **Note 24**). Let it air-dry.
4. Check chromosomes using a phase-contrast microscope. More material can be added on the same slide by repeating **step 3**.
5. Aging of slides: store slides in a vacuum chamber at 37°C overnight. Alternatively, leave at room temperature for 2–3 d.

3.4.3. Probe Labeling for FISH

DNA to be used as a probe for FISH was labeled by the nick translation method. We usually labeled the alphoid probe (p11.4) with digoxigenin and the vector probe (pPAC4) with biotin. There are slight differences in the protocol between the two labels.

3.4.3.1. BIOTIN LABELING

1. Mix 2 µL NT buffer (10X); 7.5 µL AGC nucleotide mix (0.5 m*M* stock); 2.5 µL bio-16-dUTP (1 m*M* stock); 1 µL of DNAse I (10 U/ mL in water, make fresh each time); and 500 ng DNA plus water (total volume 6 µL).
2. Mix well and pulse spin.
3. Add 1 µL polymerase I (10 U/µL) and incubate for about 90 min at 16°C.
4. Check the size of the resulting fragments by briefly running a 2-µL sample in an agarose gel. A smear of between 150–400 bp is desirable. If the smear is larger or smaller, adjust DNAse I concentration accordingly.
5. Store labeled probe at 20°C (*see* **Note 25**).

3.4.3.2. DIGOXIGENIN LABELING

The labeling procedure is identical, but using the following mixture at **step 1**:

2 µL NT buffer (10X)
7.5 µL AGC nucleotide mix (0.5 m*M* stock)
1.5 µL dTTP (0.5 m*M* stock)
1 µL dig-11-dUTP (1 m*M* stock)
1 µL DNAse I, (10 U/mL in water; make fresh each time)
500 ng DNA plus water (total volume 6 µL)

3.4.4. Hybridization

1. Incubate slides for 1 h at 37°C in 2X SSC plus 25 µg/mL RNAseA in a staining trough.
2. Wash slides briefly in 2X SSC at room temperature, and dehydrate through 70%, 90%, and 100% ethanol series for 3 min each.

3. Dry slides under vacuum. Place 70% ethanol on ice.

4. Aliquot 2–3 µL of each labeled probe, plus 0.5 µL salmon sperm (10 mg/mL stock) in individual 1.5 microfuge tubes, one per slide. Add 10 µL cold ethanol, and dry in a Speedvac machine.

5. Add 11 µL hybridization solution (*see* **Materials** section) to each tube and resuspend probe for 1 h at room temperature.

6. Warm slides in an empty staining trough in a 73°C waterbath for 5 min.

7. Denature slides in prewarmed (73°C) 70% formamide/4X SSC for exactly 2.5 min.

8. Rapidly plunge slides into ice-cold 70% ethanol and leave 3 min. Continue dehydrating through the ethanol series as above. Dry under vacuum.

9. Denature probe by placing tubes in the 73°C waterbath for 5 min and transfer quickly to ice.

10. Put 11-µL probe onto clean 22 × 22 mm coverslips and mount slides, avoiding bubbles.

11. Seal coverslips with rubber solution and place slides in metal tray inside a 37°C waterbath. Hybridize overnight.

3.4.5. Stringency Washes and Label Detection

1. Prepare blocking reagent (5% skimmed milk solution in 4X SSC). Aliquot in 1.5 mL microfuge tubes, and centrifuge at top speed for 15 min at 4°C (*see* **Note 26**).

2. The digoxigenin label can be detected with Rhodamine-conjugated antidigoxigenin. Dilute the 200 µg/mL stock 1/15 in centrifuged blocking reagent (allow 40–50 µL per slide).

3. The biotin label can be detected with FITC-conjugated avidin. Dilute the 2 mg/mL stock 1/500 in centrifuged blocking reagent (allow 40–50 µL per slide).

4. Centrifuge at top speed for 15 min at 4°C (*see* **Note 26**). Keep on ice in the dark.

5. Remove glue from slides, wash in 2X SSC at room temperature until coverslips gently fall off.

6. Wash slides 4 × 3 min in 2X SSC at 45°C.

7. Wash slides 4 × 3 min in 0.5X SSC at 60°C.

8. Transfer slides to 4X SSC at room temperature.

9. Put 50 µL centrifuged blocking reagent on 22 × 40-mm coverslips and mount on slides. Incubate for 5 min at room temperature.
10. Remove coverslip carefully (*see* **Note 27**), drain, and add 40 µL Rhodamine-conjugated antidigoxigenin to the same coverslip.
11. Mount on slides and incubate in humid chamber at 37°C for 30 min.
12. Wash in 4X SSC at room temperature until coverslips gently fall off.
13. Wash 4 × 3 min in 4X SSC/0.1% Tween 20 at 37°C.
14. Transfer slides to 4X SSC at room temperature.
15. Put 40 µL FITC-conjugated avidin onto 22 × 40-mm coverslips and mount slides.
16. Incubate in humid chamber at 37°C for 30 min.
17. Transfer slides to 4X SSC at room temperature.
18. Drain slides (do not dry) and mount in 30 µL DAPI/Vectashield with a 22 × 40-mm coverslip.
19. Carefully remove excess mountant by pressing slide between two sheets of filter paper, and seal with rubber solution. Store in the dark at 4°C.

4. Notes

1. While preparing PAC DNA using Qiagen's maxiprep kit, there are always a number of Q500 columns left by the time the solutions provided are all used up. Because these columns are the critical component of the kit, we make up our own solutions to use them.
2. Formamide is very toxic. Use with a waterbath inside a fumehood.
3. DAPI (4-6-diamidino-2-phenylindole) is light sensitive. Store at 4°C protected from the light.
4. Special care must be taken when handling large DNA molecules to minimize mechanical shearing. It is useful to keep a stock of pipet tips whose ends have been cut to widen the bore. Use these wide bore tips throughout the procedure.
5. From the indicated volume of culture, yields of 250–350 µg pTAT/7c5/BS DNA can easily be obtained.
6. It is best to add RNAse A to solution P1 just before use. However, this solution can be stored at 4°C but must be used within a month.
7. To avoid blocking the column, it is important to remove as much bacterial debris as possible. Filtering the supernatant is very effective, while the previous centrifugation helps increase the yield as

practically all of the supernatant can be recovered once the debris is compacted.

8. Avoid drying the pellet too long, otherwise the DNA will be very difficult to resuspend. Five minutes is usually enough if the supernatant is pipetted out thoroughly. The DNA pellet is often very difficult to see and it tends to be smeared along the side of the tube. For this reason, it is important to mark the orientation of the tube as it goes in the rotor.

9. For an account of the differences in MAC formation ability of circular and linear versions of a construct *see* Ebersole et al. *(11)*.

10. This may seem too much DNA for lipofection, but there are always losses during purification. In addition, large molecular weight DNA is very susceptible to mechanical shearing and degradation. Gel electrophoresis allows the separation of the coiled circular PAC molecules from the rest, and this is usually a very small proportion of the total DNA loaded. A smear usually appears at the area where the linear form would appear because nicked molecules accumulate here, and this may complicate the separation of digestion-linearized molecules from randomly nicked ones. This can be minimized by using fresh DNA preparations, and of course, by exercising due care when manipulating these large molecules.

11. Several comb teeth can be taped together to give a well of the appropriate volume. We normally use about 300 μL to hold 30 μg DNA.

12. It is useful to leave out the edges of the large band to be stained with the rest because it helps locate the DNA bands.

13. Fifteen minutes is usually enough. Mix occasionally by inversion. After the agarose is completely molten, pulse spin the tubes and place at 65°C for a couple of minutes to make sure all of the agarose is at the bottom of the tube.

14. Although 90 min is normally enough, it is often safer to incubate for up to 2 or 2.5 h because if the digestion is not complete the whole procedure would have to be repeated.

15. The DNA can be safely kept at 4°C in isopropanol for 2–3 d if it is not practical to proceed with the lipofection immediately.

16. HT1080 cells double every 18–20 h approximately. Incubate at 37°C with 5% CO_2.

17. Although 1–2 μg are recommended, as little as 100–150 ng have been successfuly lipofected to induce MAC formation.

18. It is important that both signals can be colocalized in a putative MAC to be sure that it is indeed derived from the input DNA. This is valid for routine tests. Probe p11.4 is specific for the alphoid arrays on chromosomes 13 and 21 at very high stringency. However, it is common to obtain positive signals on other centromeres if the stringency is lowered a little. In addition to this FISH test, further probes should be employed to ascertain whether the putative MAC has incorporated sequences from the host's genome. The presence of an active centromere in the putative MAC cannot be inferred solely from the presence of centromeric alphoid DNA. It is necessary to use immunocytochemistry techniques to detect centromeric proteins such as CENP-C and CENP-E, which are known to be present only at active centromeres. The choice of probes for FISH and the immunocytochemistry method are illustrated in Ebersole et al. *(11)*.

19. Mitotic cells can easily be distinguished as they become round. Monitor the appearance of the cells to judge the effectiveness of the colcemid treatment.

20. In a standard benchtop centrifuge with a maximum radius of 18 cm. this corresponds to about 1000 rpm.

21. Timing is very important. Too short, and the chromosomes will appear clumped. Too long and the cells will burst easily making complete metaphases difficult to find.

22. Usually we proceed to make slides to check the quality of the chromosome preparations. These are only stored for later use if they are good enough.

23. Keeping the tubes chilled helps improve significantly the quality of the spreads.

24. Although some material is lost in this procedure, what remains is very clean and metaphase spreads are usually devoid of cytoplasm. This is optimal for FISH.

25. Unincorporated nucleotides can be removed by chromatography, but in our experience this step is not always necessary although it can result in cleaner slides.

26. Centrifugation of any solution containing skimmed milk helps remove solids in suspension. This reduces nonspecific background and results in cleaner and clearer FISH signals.

27. Flicking the slide sharply edgeways is enough to displace the coverslip slightly, and can then be easily removed.

Acknowledgments

We would like to thank our colleagues, present and past, for their input in the development of these methods: Elma Clark, Andrew Ross, Niolette McGill and, in particular, Tom Ebersole and Brenda Grimes, who originally developed this system.

References

1. Newlon, C. S. (1988) Yeast chromosome replication and segregation. *FEMS Microbiol. Rev.* **52,** 568–601.
2. Farr, C.J., Stevanovic, M., Thomson, E. J., Goodfellow, P. N., and Cooke, H. J. (1992) Telomere-associated chromosome fragmentation: application in genome manipulation and analysis. *Nat. Genet.* **2,** 275–282.
3. Auriche, C., Donini, P., and Ascenzioni F. (2001) Molecular and cytological analysis of a 5.5 Mb minichromosome. *EMBO Rep. Feb.* **2,** 102–107.
4. Heller, R., Brown, K. E., Burgtorf C., and Brown W. R. (1996) Minichromosomes derived from the human Y chromosome by telomere directed chromosome breakage. *Proc. Natl. Acad. Sci. USA* **93,** 7125–7130.
5. Shen, M. H., Mee, P. J., Nichols, J., Yang, J., Brook, F., Gardner, R. L., et al. (2000) A structural defined mini-chromosome vector for the mouse germline. *Curr. Biol.* **10,** 31–34.
6. Murphy, T. D. and Karpen G. H. (1995) Localization of centromere function in a Drosophila minichromosome. *Cell* **82,** 599–609.
7. Patnaik, P. K., Axelrod, N., Van der Ploeg, L. H., and Cross, G. A. (1996) Artificial linear mini-chromosomes for *Trypanosoma brucei. Nucleic Acids Res.* **24,** 668–675.
8. Harrington, J. J., Van Bokkelen, G., Mays, R.W., Gustashaw, K., and Willard, H. F. (1997) Formation of de novo centromeres and construction of first-generation human artificial microchromosomes. *Nat. Genet.* **15,** 345–355.
9. Ikeno M., Grimes, B., Okazaki, T., Nakano, M., Saitoh, K., Hoshino, H., et al. (1998) Construction of YAC-based mammalian artificial chromosomes. *Nat. Biotechnol.* **16,** 431–439.

10. Henning K.A., Novotny E. A., Compton S. T., Guan X. Y., Liu P. P., and Ashlock M. A. (1999) Human artificial chromosomes generated by modification of a yeast artificial chromosome containing both human α-satellite and single-copy DNA sequences. *Proc. Natl. Acad. Sci. USA* **96,** 592–597.

11. Ebersole T. A., Ross A., Clark E., McGill, N., Schindelhauer D., Cooke, H., et al. (2000) Mammalian artificial chromosome formation from circular alphoid input DNA does not require telomere repeats. *Hum. Mol. Genet.* **9,** 1623–1631.

12. Ten Hagen, K. G., Gilbert, D. M., Willard, H. F., and Cohen, S. N. (1990) Replication timing of DNA sequences asociated with human centromeres and telomeres. *Mol. Cell. Biol.* **10,** 6348–6355.

13. Frengen, E., Zhao, B., Howe, S., Weichenhan D., Osoegawa, K., Gjernes, E., et al. (2000) Modular bacterial artificial chromosome vectors for transfer of large inserts into mammalian cells. *Genomics* **68,** 118–126.

11

Use of Natural and Artificial Chromosome Vectors for Animal Transgenesis

Yoshimi Kuroiwa, Kazuma Tomizuka, and Isao Ishida

1. Introduction

1.1. Generation of Trans-Chromosomic (Tc) Mice by Using Natural Human Chromosome Fragments (hCFs)

The ability to use hCFs as vectors for introducing large stretches of human DNA into mice was first demonstrated in 1997 *(1)*. Transferred hCFs were stably maintained as an extra chromosome in the somatic cells of mice, and the human genes included in them were expressed under proper tissue-specific regulation. In some cases they could be transmitted through the germline, resulting in the establishment of novel mouse strains (*trans*-chromosomic [Tc] mice) containing a heritable hCF *(1,2)*. Thus, such an approach using chromosome vectors for animal transgenesis has been thought to be useful for overcoming size constraints of cloned transgenes in conventional techniques and facilitate functional studies of human genome in the postsequencing era.

From: *Methods in Molecular Biology, Vol. 240:*
Mammalian Artificial Chromosomes: Methods and Protocols
Edited by: V. Sgaramella and S. Eridani © Humana Press Inc., Totowa, NJ

Compared with conventional techniques using cloned transgenes, the chromosome vector system has several advantages: (1) whole genomic sequences, including all the introns and essential regulatory elements, can be used as transgenes, which should confer correct and physiologically controlled expression of transgenes in vivo; (2) very large genes, gene clusters, and specific chromosomal regions, which cannot be cloned as contiguous DNA fragments by cloning techniques, can be introduced; (3) chromosome vectors are maintained as a single-copy extra chromosome in the host cells, which may prevent toxic overexpression or gene silencing *(3)* caused by multiple copy insertion of transgenes; and (4) there is no risk for insertional mutagenesis because they are freely segregating from host chromosomes.

Our strategy to introduce chromosome vectors into mice is outlined in **Fig. 1**. We used microcell-mediated chromosome transfer (MMCT) to introduce a specific human chromosome into mouse embryonic stem (ES) cells *(1)*.

For this purpose, we constructed a new library of human–mouse A9 monochromosomal hybrids containing human chromosomes derived from normal embryonic fibroblasts *(1)*. The library comprises approx 700 independent hybrid clones, each of which contains a human chromosome, randomly tagged with pSTneoB *(1)*, suitable for conferring G418-resistance to mouse ES cells. This can be screened by polymerase chain reaction (PCR) to identify A9 hybrids containing the gene or chromosomal region to be introduced into mouse ES cells.

To date, we successfully produced chimaeric mice from microcell hybrid ES (MH[ES]) cells containing the hCF derived from hChr2, 4, 6, 7, 11, 14, 21, or 22. In addition, germline transmission of transferred hCF was observed in chimaeras produced from MH(ES) cells containing the hChr2, 7, 14, or 21-derived hCF *(1,2,4,5)*.

1.2. Construction of Human Artificial Chromosomes (HACs)

As described, we successfully used hCFs as vectors for introducing large human genomic region into mice. However, one major

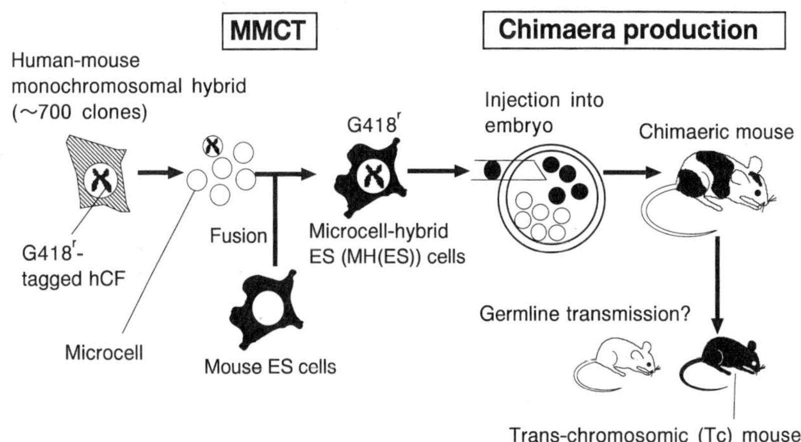

Fig. 1. Schematic diagram showing the construction of MH(ES) cells to produce chimaeric mice expressing human genes on the transferred hCFs. G418^r: G418-resistant.

problem of hCF vectors has been that they are structurally undefined *(6,7)*. The hCFs are usually generated as a consequence of accidental fragmentation of intact chromosome during the chromosome transfer process and run a risk to contain unrelated, deleterious genes in addition to the genes of interest. There have been some limitations in using structurally undefined hCFs because (1) mitotic stability of hCF varies among individual hCFs *(8)*; (2) gene expression from the introduced hCFs may affect mouse development; and (3) successful germline transmission in mice has been observed in limited kind of hCFs. To resolve above problems, we have recently developed a chromosome cloning system where an unstable hCF can be cloned into a stable and germline-transmittable HAC vector by combination of Cre/loxP-mediated chromosomal translocation with telomere-directed chromosomal truncation in homologous recombination-proficient chicken DT40 cells *(9)*. The complete human genome sequence, which has now been available *(10)*, should facilitate the minimization of HACs containing only desired chromosomal regions.

As a basal human minichromosome vector, we chose to use the hChr14-derived hCF(SC20) because it was stably maintained and germline-transmitted in mice *(2)*. In the DT40 cells, a loxP sequence was integrated at *RNR2* locus on the hCF(SC20) by homologous recombination (SC20 vector), into which a genomic DNA region of interest could be cloned by Cre/loxP recombination.

For this purpose, the human chromosome containing a locus of interest can be modified as follows: in the DT40 cells, (1) human telomeric repeat $(TTAGGG)_n$ is integrated by homologous recombination at a locus just telomeric to the locus of interest, which leads to chromosome truncation at the integration site of the telomeric repeat; (2) a loxP sequence is integrated by homologous recombination at a locus just centromeric to the locus of interest. The DT40 cells containing the SC20 vector and the DT40 cells containing the modified human chromosome (hChr) fragment of interest are fused, resulting in the DT40 hybrids containing both of hChr fragments. To induce chromosomal translocation between the SC20 vector and the hChr fragment of interest by Cre/loxP-mediated recombination, a Cre recombinase–expression vector is introduced into the DT40 hybrids. The generation of the HAC where the locus of interest defined by the loxP-integration and telomere-truncation sites is cloned into the loxP-cloning site on the SC20 vector can be confirmed by PCR and fluorescence *in situ* hybridization (FISH) analyses (outlined in **Fig. 2**). Once the HAC carrying the locus of interest is generated, it can be transferred to various types of cells where phenotypic analyses can be performed. When the HAC is transferred into mouse ES cells, chimeric mice carrying the HAC can be generated for its functional analysis in vivo *(8)*.

By using the chromosome-cloning system, various hChr regions could be cloned into the SC20 minichromosome vector, irrespective of its size, and introduced into mice stably, allowing human genes of interest to be functionally expressed in vivo. In our HACs, the chromosome inserts flanked by the loxP-integration and telomere-truncation sites are structurally determined by the information of the recently published human genome sequence *(10)*. The availability of structurally defined HAC vectors would be of great value in the construction of animals carrying human genetic elements to

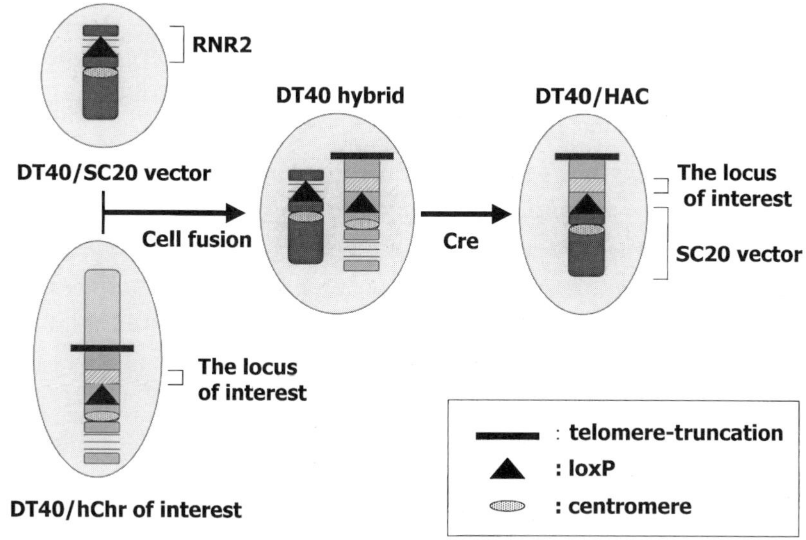

Fig. 2. Schematic diagram showing the chromosome cloning approach.

model specific diseases or the production of various therapeutic products for industrial purposes.

Another potential application of the HAC vector may be the use in the treatment of human genetic diseases. Indeed, the advantages of chromosome vector system in animal transgenesis mentioned above are also desirable characteristics for therapeutic vector to overcome various problems in exiting viral and nonviral vector systems. Although the inefficient delivery of a large chromosome molecule into the cells remains to be overcome our trans-chromosomic mice may provide the useful model system for in vivo assessment of the function and behavior of chromosome-based therapeutic vectors. As for the hChr14-derived hCF(SC20) minichromosome vector, we are in the process of minimizing it in size so that it will no longer contain extra human genes on it, aiming at using it as a vector for human gene therapy. For example, *dystrophin* is the largest gene (2.4 mb) in humans and responsible for Duchenne muscular dystrophy. Because of its large size, the entire *dystrophin* locus has never been cloned even by using YAC vectors *(11)*. However, by using our HAC system, it can be cloned and the HAC carrying entire *dystrophin* locus could be evaluated for its usefulness as a therapeutic vector for Duchenne muscular dystrophy in a model animal, such as mdx mice *(12)*.

Our studies using Tc mice with hCFs or HACs have demonstrated the utility of Tc technology as a complementary approach to conventional transgenic techniques using the cloned DNA fragments *(1,2,9)*. Here, we summarize current protocols for this novel approach, which should allow for large-scale manipulation of mammalian genome to generate genetically modified animals useful for laboratory and industrial uses.

2. Materials

2.1. Construction of MH(ES) Cells

1. Recipient mouse ES cell line: TT2F *(1)* (*see* **Note 1**). Detailed procedures for ES cell culture are described in **ref. 13**.
2. Donor A9 hybrid cell line (G418[r]) selected from the library (*see* **Notes 2** and **3**).
3. Dulbecco's modified Eagle's medium (DMEM).
4. ES cell qualified fetal bovine serum (FBS; Gibco BRL).
5. Colcemid: Demecolcine (Sigma).
6. A9 hybrid medium: DMEM with 10% fetal calf serum (FCS).
7. Microcell formation medium: DMEM with 20% FBS and 0.05 μg/mL Colcemid.
8. ES medium: DMEM (4.5% [w/v] glucose) with 18% FCS, nonessential amino acids solution supplement (100X, Gibco BRL) and 10^3 U/mL leukemia inhibitory factor (ESGRO, Amrad).
9. Cytochalasin B (Sigma): Stock solutions of 2 mg/mL in dimethyl sulfoxide are stable indefinitely at 4°C if protected from light.
10. Cytochalasin B medium: DMEM with 10 μg/mL of cytocharasin B.
11. Primary fibroblast: Neo-resistant primary fibroblast (Lifetech Oriental, Japan).
12. Centrifuge flask: Nunc152094 (25 cm^2).
13. Centrifuge tube: Nalgene3141 (450 mL).
14. PEG solution (1:1.4): Dissolve 1 g of PEG1000 in 1.2 mL DMEM and 0.2 mL of dimethyl sulfoxide.

2.2. Construction of HACs

1. RPMI1640 medium.
2. DT40 culture medium: RPMI1640 with 10 % FBS, 1% chicken serum, 10^{-4} *M* 2-mercaptethanol.
3. Tris/ethylenediamine tetra-acetic acid (pH 8) buffer (TE).
4. Capaciter discharge device: Bio-Rad gene pulser II.
5. DMEM.
6. Polyethylene glycol (PEG) 1500 (Sigma).

3. Methods

3.1. Construction of MH(ES) Cells

1. Donor A9 hybrid cells are cultured in 24 centrifuge flasks until the cell density reaches 70–80% confluence.
2. Replace the medium with a microcell formation medium, followed by incubation of the cells for 2 d at 37°C.
3. After aspiration of the medium, fill the centrifuge flask by cytochalasin B medium preincubated at 37°C. Screw up the cap of flask tightly.
4. Insert the flask into an 500-mL centrifuge tube containing 300 mL of sterilized water preincubated at 37°C. Put the centrifuge tube into the rotor. The bottom of flasks should be inner side of the rotor. Centrifuge the tubes at 8000 rpm for 1 h at 34°C.
5. Remove cytochalasin B medium and resuspend the precipitate of microcells in 2 mL serum-free DMEM. Pool the cell suspension in 50-mL Falcon tube.
6. Purify relatively small microcells by successive passage through syringe-top 8-, 5-, and 3-μm filters (Costar, NUCLEPORE). Centrifuge filtered microcell suspension for 7 min at 1500 rpm.
7. Disperse 1×10^{7} of logarithmically growing mouse ES cells with trypsin and wash three times with serum-free DMEM. Resuspend the cells in 5 mL DMEM.
8. Overlay the ES cell suspension on the centrifuged microcells, followed by the centrifuge for 7 min at 1250 rpm.
9. Remove supernatant. The precipitate should be tapped vigorously to disperse any clumps of ES cells and microcells.

10. Slowly drip the 0.5 mL of PEG solution (1:1.4) preincubated at 37°C while gently dispersing the cells for 90 s. Immediately add 10 mL of serum-free DMEM in a dropwise fashion, with gentle swirling over 1 min.
11. Centrifuge fusion mixture at 1250 rpm for 7 min. Aspirate the media.
12. After resuspension of the precipitate in 20 mL ES medium, transfer the cell suspension into two 100-mm dishes with neoresistant feeder layer. Incubate for 36 h at 37°C.
13. Replace the ES medium with a medium supplemented with 200 μg/mL o G418. Change the media every day.
14. Drug-resistant ES colonies should be visible in 1 wk.

Structural analysis of transferred hCF in the resultant drug-resistant ES clones must be performed by PCR and FISH to determine whether chromosome fragmentation has occurred during MMCT. Typically, 10–40% of drug-resistant, MH(ES) clones contain the hCF associated with further deletions. In addition to MH(ES) cell lines retaining an structurally unchanged hCF, those retaining fragmented hCFs with the gene of interest can also be used in the following experiments.

3.2. Construction of a HAC Carrying the Locus of Interest in the DT40 Cells

3.2.1. Modification of hChr Containing a Locus of Interest

The hChr or its fragment containing the locus of interest can be transferred from the A9 cells to DT40 cells by the MMCT technique. In the DT40 cells, the hChr can be modified by combination of telomere-directed chromosome truncation with a loxP integration, by which the locus of interest can be defined and flanked by the telomere-truncation and loxP-integration sites. The defined locus is subjected to cloning into the SC20 vector to generate the HAC.

3.2.1.1. Telomere-Directed Chromosome Truncation

To truncate the hChr, a targeting vector is required to integrate human telomeric repeat to the targeted locus. The basal structure of the vector we routinely use is shown in **Fig. 3**. This vector is com-

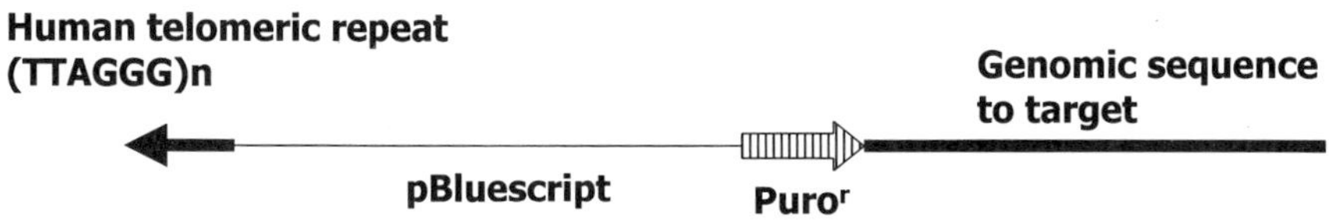

Fig. 3. Structure of the vector for telomere-directed chromosome truncation.

posed of the genomic sequence to target, puromycin-resistant gene for drug selection and human telomeric repeat $(TTAGGG)_n$. We usually obtain the genomic sequence ranging from 5 to 10 kb by PCR-based amplification and the amplified DNA is subcloned into the pTELPuro basal vector. The vector is electroporated into the DT40 cells containing the hChr and then transfectants are selected by puromycin for several weeks. After selection, genomic DNA is extracted from picked up individual colony and subjected to PCR screening on the basis of a series of deletions of STS markers telomeric to the truncation site. Furthermore, the truncation of the hChr can be visually confirmed by FISH analysis.

So far, we have succeeded in targeted truncation of hCh2, 6, 7, 22, and X at a specific locus and found that the truncation occurs in all the targeted clones where the vector is homologously integrated at the target locus, suggesting that telomere-truncation can efficiently occur in DT40 cells.

1. Seed 45 mL of nonselective DT40 medium with logarithmically growing DT40 cells containing the hChr of interest prior the day of transfection.
2. Linearize 25–30 μg/transfection of targeting plasmid containing telomeric repeats with the appropriate restriction enzyme. After digestion, restriction enzyme reactions should be extracted with phenol/$CHCl_3$, precipitated, and resuspended to 1 μg/μL in TE.
3. Count the DT40 cells and pellet 1×10^7 cells/transfection by centrifugation. Rinse the cell pellet with serum-free RPMI 1640 once and then resuspend at 1×10^7 cells/0.5 mL of serum-free RPMI1640.
4. For each transfection, transfer 0.5 mL of the cell suspension to a sterile electroporation cuvet. Add 25–30 μg of the linearized DNA to the cuvet; mix gently by pipetting and place it at room temperature for 10 min.

5. Electroporate the DT40 cells at 25 μF and 550 V using a Bio-Rad gene pulser II. The time constant should be 0.7–0.8 ms.

6. After a 10-min incubation at room temperature after electroporation, transfer the cells to a T-25 flask containing 20 mL of nonselective DT40 medium and then incubate them for 24 h at 37°C.

7. Resuspend each transfection in selective DT40 medium supplemented with 0.3 μg/mL of puromycin and divide to 10–20 × 96-well microtiter plates with 0.2 mL/well. Transfectant colonies should be visible in 10–14 d.

8. Pick up transfectant colonies individually and then transfer to 24-well plates. A few days after, each transfectant can be transferred to 6-well plates, from part of which genomic DNA can be extracted to screen the occurrence of targeted chromosomal truncation.

First, we routinely use PCR screening to identify homologous recombination using the junction primer set as shown in **Fig. 3**. After that, the occurrence of the truncation can be determined by STS mapping around the targeted locus; all markers telomeric to the locus should be deleted whereas all markers centromeric to the locus should remain. Furthermore, the chromosomal truncation can be visually confirmed by FISH analysis.

3.2.1.2. Lox P integration into the Truncated hChr

The DT40 cells containing the truncated hChr can be sequentially transfected with a targeting vector containing a loxP sequence. For this purpose, we typically use a loxP-targeting vector, shown in **Fig. 4**. This vector consists of the genomic sequence to target, hygromycin B-resistant gene for drug selection, a loxP sequence, and PGK promoter. We usually obtain the genomic sequence ranging from 7 to 10 kb by PCR-based amplification and the amplified DNA is subcloned into the ploxPHyg basal vector. The vector is electroporated into the DT40 cells containing the truncated hChr and then transfectants are selected by hygromycin B for several weeks. After selection, genomic DNA is extracted from picked up individual colony and subjected to PCR screening to identify homologous recombination using the junction primer set as shown in **Fig. 4**.

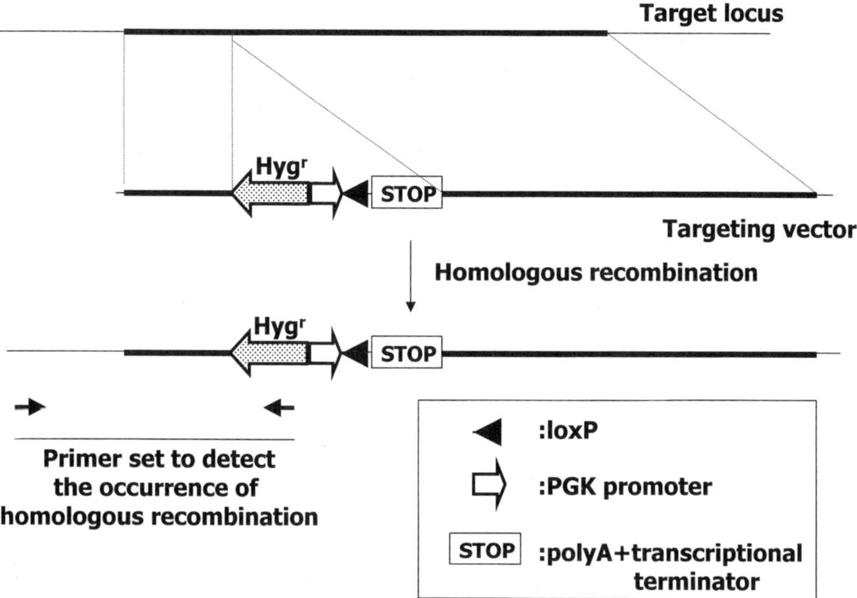

Fig. 4. Structure of the vector for LoxP site integration and the detection of homologous recombination by PCR.

1. Seed 45 mL of nonselective DT40 medium with logarithmically growing DT40 cells containing the truncated hChr prior the day of transfection.
2. Linearize 25–30 µg/transfection of a targeting plasmid containing a loxP sequence with the appropriate restriction enzyme. After digestion, restriction enzyme reactions should be extracted with phenol/$CHCl_3$, precipitated and resuspended to 1 µg/µL in TE.
3. Count the DT40 cells and pellet 1×10^7 cells/transfection by centrifugation. Rinse the cell pellet with serum-free RPMI 1640 once and then resuspend at 1×10^7 cells/0.5 mL of serum-free RPMI1640.
4. For each transfection, transfer 0.5 mL of the cell suspension to a sterile electroporation cuvet. Add 25–30 µg of the linearized DNA to the cuvet; mix gently by pipetting and place it at room temperature for 10 min.
5. Electroporate the DT40 cells at 25 µF and 550 V using a Bio-Rad gene pulser II. The time constant should be 0.7–0.8 ms.

6. After a 10-min incubation at room temperature after electroporation, transfer the cells to a T-25 flask containing 20 mL nonselective DT40 medium and then incubate them for 24 h at 37°C.

7. Resuspend each transfection in selective DT40 medium supplemented with 1 mg/mL of hygromycin B and divide to 10–20 × 96-well microtiter plates with 0.2 mL/well. Transfectant colonies should be visible in 10–14 d.

8. Pick up transfectant colonies individually and then transfer to 24-well plates. A few days after, each transfectant can be transferred to 6-well plates, from part of which genomic DNA can be extracted to screen the occurrence of targeted integration of the loxP sequence.

3.2.2. Whole Cell Fusion

1. Seed 45 mL nonselective DT40 medium with logarithmically growing DT40 clone containing the modified hChr and one containing the SC20 vector, individually, prior the day of fusion.

2. Count the each DT40 clone and mix 1–2×10^7 cells of the DT40 clone containing the modified hChr with those of the DT40 clone containing the SC20 vector, followed by centrifugation. Rinse the cell pellet with serum-free DMEM.

3. After aspiration of the medium completely, drip 0.5 mL PEG on the pellet and stir for 2 min.

4. Immediately add 1 mL serum-free DMEM, with gentle stirring over 1 min. After that, add 9 mL of serum free-DMEM, with gentle stirring over 2 min.

5. After a 10-min incubation at 37°C, centrifuge the fusion mixture and then transfer the cells to a T-25 flask containing 20 mL nonselective DT40 medium. Incubate them for 24 h at 37°C.

6. Resuspend the fusant cells in selective DT40 medium supplemented with 1 mg/mL hygromycin B and 10 µg/mL blasticidin-S, and then divide to 4 × 24-well plates with 0.5 mL/well. Fusant colonies should be visible in 15–20 d.

7. Pick up fusant colonies individually and then transfer to 6-well plates, from part of which genomic DNA can be extracted to see if the fusant clone contains both the SC20 vector and the modified hChr fragment.

We typically obtain five fusant clones per cell fusion. The retention of both the SC20 vector and the modified hChr fragment can be easily determined by PCR analysis using STS primers specific to the chromosomes, as well as by FISH analysis using painting probes specific to the chromosomes.

3.2.3. Introduction of Cre Recombinase to Induce Chromosomal Translocation

Once the fusant DT40 hybrids retaining both the SC20 vector and the modified hChr fragment are obtained, Cre recombinase can be introduced into the hybrids to induce chromosomal translocation between the two loxP sites on the hChrs. As one way to introduce Cre recombinase, its transient expression can be considered, but the frequency of chromosomal translocation between nonhomologous chromosomes is reported to be very low in mouse ES cells *(14)*. Furthermore, in the case of translocation between exogenous nonhomologous hChrs in nonhuman cells (chicken DT40 cells), its frequency is suspected to be a little lower than that of endogenous nonhomologous chromosomes. Therefore, we routinely take a strategy to stably express Cre recombinase, instead of transient expression. For this purpose, we use a modified Cre recombinase expression vector (derived from pBS185) containing a hisD-resistant gene to select stable transfectants. After selection by hisD, occurrence of the translocation can be confirmed by PCR using the primers shown in **Fig. 5** and FISH analyses using chromosome-specific probes.

1. Seed 45 mL selective DT40 medium containing 1 mg/mL of hygromycin B and 10 µg/mL of bsr with logarithmically growing the DT40 hybrid cells retaining the SC20 vector and the modified hChr fragment prior the day of transfection.
2. Linearize 30 µg/transfection of a Cre-expression vector with the appropriate restriction enzyme. After digestion, restriction enzyme reactions should be extracted with phenol/CHCl$_3$, precipitated and resuspended to 1 µg/µL in TE.

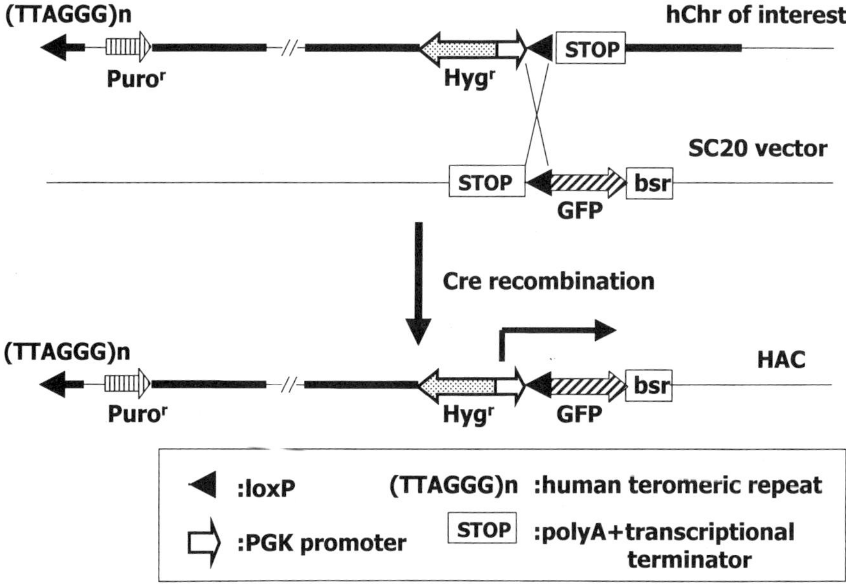

Fig. 5. Detection of interchromosomal translocation by GFP expression.

3. Count the DT40 cells and pellet 1×10^7 cells/transfection by centrifugation. Rinse the cell pellet with serum-free RPMI 1640 once and then resuspend at 1×10^7 cells/0.5 mL serum-free RPMI 1640.

4. For each transfection, transfer 0.5 mL cell suspension to a sterile electroporation cuvet. Add 30 μg of the linearized DNA to the cuvet; mix gently by pipetting and place it at room temperature for 10 min.

5. Electroporate the DT40 cells at 25 μF and 550 V using a Bio-Rad gene pulser II. The time constant should be 0.7–0.8 ms.

6. After a 10-min incubation at room temperature after electroporation, transfer the cells to a T-25 flask containing 20 mL of nonselective DT40 medium and then incubate them for 24 h at 37°C.

7. Resuspend each transfection in selective DT40 medium supplemented with 5 mg/mL of hisD and divide to a single 12-well microtiter plate with 1 mL/well. Transfectant colonies should be visible in 10–14 d.

8. Transfer to 2×6-well plates, from a half amount of which genomic DNA can be extracted to screen the occurrence of the chromosomal translocation.

Once the occurrence of the chromosomal translocation is confirmed, the DT40 cells containing the translocated chromosome fragment, that is HAC, need to be isolated by positive selection, such as drug resistance and indicator gene expression. To isolate the cells, we routinely use fluorescence-activated cell sorting on the basis of the expression of green fluorescent protein (GFP) gene.

As shown in **Fig. 5**, in our system, the GFP gene is engineered to be transcribed by PGK promoter as a result of occurrence of the chromosomal translocation. By FACS sorting of the GFP-positive cells, the cells containing the HAC can be easily isolated. So far, we have succeeded in construction of six HACs in which several hChr regions (hChr2, 6, 7, and 22) of interest are cloned into the SC20 vector.

4. Notes

1. ES cell line: It has been recognized that male meiosis is more sensitive to the presence of an unpaired, univalent chromosome *(15)* than female meiosis. We therefore believe that the female mouse ES cells (39, XO) may be more suitable as a recipient for the chromosome transfer than male ES cells (40, XY) usually employed in knockout studies *(1,2)*. We routinely use a ES cell line, TT2F (39, XO) *(1)*, a derivative of TT2 (40, XY) *(1)*, which spontaneously lost the Y chromosome.

2. Screening of monochromosomal hybrid library: For example, PCR analyses using primer pairs specific for human Igκ gene (2q12) resulted in 16 Igκ-positive hybrid clones. Detailed analyses showed that 4 of 16 clones contained an intact human chromosome 2. The G418[r] marker was found to be inserted in the unique site of human chromosome 2 in each of these 4 hybrid clones by fluorescence *in situ* hybridization (FISH), which demonstrated the randomness of this library. Another library containing over 700 hybrid clones are reported by Inoue et al. and also available *(16)*.

3. Selection of hCF: The size of hCF is the most important point to be considered in the selection of hCF for the use in generating Tc mice. Our previous studies suggested that the use of small hCFs may increase the probability of successful chimera production from

microcell hybrid ES (MH[ES]) cells and germline transmission of the hCFs *(2)*. It should prevent developmental abnormalities in chimeric mice, caused by overexpression of unrelated, deleterious genes contained in the transferred hCFs or meiotic arrest caused by the presence of unpaired, univalent chromosome. Therefore, a panel of A9 hybrid clones, obtained by screening the library by PCR primers specific for the gene of interest, should then be subjected to more detailed PCR and FISH analysis to find a hCF as small as possible. In addition, the presence of a drug-resistant marker on the hCF in selected A9 hybrid clone should be confirmed by FISH.

4. Chimera production: Standard procedure for chimera production as described in **ref.** *13* can be applyed to that from MH(ES) cells. In the case of TT2F cell line, injection of MH(ES) cells into eight cell embryos prepared from albino ICR mice should result in chimeric mice with high chimerism.

5. Stability of hCFs: It has been suggested that each hCF may have an intrinsic level of mitotic and miotic stability *(8)*. The stability of hCFs in MH(ES) cells is therefore another crucial point to be considered in the choice of hCF for the use in generating Tc mice. Long-term stability of hCFs should be tested by culturing the MH(ES) cells for at least 30 d under the nonselective condition. Our previous studies *(2,9)* revealed that the hChr14-derived hCF (hCF[SC20]) was highly stable (<0.1% loss/doubling) in contrast to the hChr2-derived hCF (hCF[2-W23]) (3.2% loss/doubling), hChr22-derived hCF (hCF[22]) (>5% loss/doubling) and the hChrY-derived hCF reported by Shen et al. *(17)*. Similar results were obtained in various types of cell lines from different species and in the somatic and germ cells of Tc mice *(8)*. Although the use of stable hCFs may be desirable to ensure the expression of introduced genes in the somatic cells of mice and efficient germline transmission of the hCFs, in practical, MH(ES) cell lines containing a hCF whose loss rate is less than 5%/doubling can be used for the chimera production. Indeed, we successfully generated chimeric and Tc mice from MH(ES) cells retaining the hCF(2-W23) (3.2% loss/doubling, *see* above), in which substantial fraction of somatic cells retained the hCFs.

6. Expression of introduced genes and phanotypic analysis: Phenotypic changes caused by expression of genes residing in the transferred chromosome can also be addressed in chimeric mice. For example, chimeric mice containing a human chromosome 21 showed a high correlation between retention of the transferred chromosome in the

brain and impairment in learning or emotional behavior *(4)*. Functional expression of introduced genes was extensively investigated in the chimeric and Tc mice containing the hCF(2-W23) and hCF(SC20), each of which include immunoglobulin (Ig) kappa light chain (2 Mb) and heavy chain (1.5 Mb) locus, respectively *(1,2)*. The results of structural analysis of human Ig mRNA and serum expression of Ig proteins suggested that these large and complex loci were properly expressed to reconstitute the diverse and functional repertoire of human Ig in mice *(1,2)*. Furthermore, the introduction of hCF(SC20) into endogenous IgH-knockout strain, in which functional B lymphocytes and Ig production are absent, by mating resulted in the rescue of its defects, indicating that the stability and functionality of the hCF(SC20) is likely to be sufficient for restoration of B cells in adult mice *(2)*. Thus, transmittable hCFs can be used for humanization of over megabase-sized, complex loci in mice.

7. Germline transmission of hCFs: Germline transmission of hCF vectors to the next generation through the germline was surprising results because it was thought that many problems remained to be solved for successful transmission of transferred foreign chromosomes *(1)*. Indeed, the possibility that an extra human chromosome may inhibit the differentiation of MH(ES) cells into functional germ cells is suggested by observed sterility in some male chimaeras containing the hChr14-derivatives *(1)*. Transmission efficiencies of hCFs may be affected by various factors, including mitotic stability of hCFs, meiotic arrest by the presence of univalent chromosome and expression of genes residing in the transferred chromosome during germ cell development. In the case of hCF(2-W23), observed transmission efficiency was 7% in male and 25% in female *(2)*. The expected transmission efficiency in chimaeras and Tc mice hemizygous for the hCF is 50% when mitotic stability of the hCF is perfect and it can be properly segregate in miosis. Considering that 62 and 25 cell divisions are required to generation of mature sperm and oocyte, respectively *(18)*, in mice, expected efficiencies in Tc(W23), deduced from the mitotic loss rate in MH(ES) cells (3.2% loss/doubling), are 7% in male and 22% in female. These efficiencies are consistent to observed ones (*see* above). Thus, it was suggested that the transmission efficiency was mainly affected by mitotic stability in Tc mice containing the hCF(2-W23). On the other hand, the results from another transmittable hCF, hCF(SC20), indicated the involvement of other factors than its mitotic stability *(2)*.

Acknowledgments

These methods were developed in the context of a collaboration with Drs. Mitsuo Oshimura, Tottori University, Japan, and Kazunori Hanaoka, Kitasato University, Japan.

References

1. Tomizuka, K., Yoshida, H., Uejima, H., Kugoh, H., Sato, K., Ohguma, A., et al. (1997) Functional expression and germLine transmission of a human chromosome fragment in chimaeric mice. *Nat. Genet.* **16,** 133–143.
2. Tomizuka, K., Shinohara, T., Yoshida, H., Uejima, H., Ohguma, A., Tanaka, S., et al. (2000) Double trans-chromosomic mice: maintenance of two individual human chromosome fragments containing Ig heavy and kappa loci and expression of fully human antibodies. *Proc. Natl. Acad. Sci. USA* **97,** 722–727.
3. Garrick, D., Fiering, S., Martin, D. I., and Whitelaw, E. (1998) Repeat-induced gene silencing in mammals. *Nat. Genet.* **18,** 56–59.
4. Shinohara, T., Tomizuka, K., Miyabara, S., Takehara, S., Kazuki, Y., Inoue, J., Katoh, M., et al. (2001) Mice containing a human chromosome 21 model behavioral impairment and cardiac anomalies of Down's syndrome. *Hum. Mol. Genet.* **10,** 1163–1175.
5. Kazuki, Y., Shinohara, T., Tomizuka, K., Katoh, M., Ohguma, A., Ishida, I., et al. (2001) Germline transmission of a transferred human chromosome 21 fragment in transchromosomal mice. *J. Hum. Genet.* **46,** 600–603.
6. Grimes, B. and Cooke, H. (1998) Engineering mammalian chromosomes. *Hum Mol Genet.* **7,** 1635–1640.
7. Brown, W. R. A., Mee, P. J., and Shen, M. H. (1998) Artificial chromosomes: Ideal vectors? *Trends Biochem.* **18,** 218–223.
8. Shinohara, T., Tomizuka, K., Takehara, S., Yamauchi, K., Katoh, M., Ohguma, A., et al. (2000) Stability of transferred human chromosome fragments in cultured cells and in mice. *Chrom. Res.* **8,** 713–725.
9. Kuroiwa, Y., Tomizuka, K., Shinohara, T., Kazuki, Y., Yoshida, H., Ohguma, A., et al. (2000) Manipulation of human minichromosomes to carry greater than megabase-sized chromosome inserts. *Nat. Biotechnol.* **18,** 1086–1090.

10. International Human Genome Sequencing Consortium. (2001) A physical map of the human genome. *Nature* **409,** 860–921.

11. Dunnen, J. T. D., et al. (1992) Recombination of the 2.4Mb human DMD-gene by homologous YAC recombination. *Hum. Mol. Genet.* **1,** 19–28.

12. Ryder-Cook, A. S., Sicinski, P., Thomas, K., Davies, K. E., Worton, R. G., Barnard, E. A., et al. (1988) Localization of the mdx mutation within the mouse dystrophin gene. *EMBO J.* **7,** 3017–3021.

13. Hogan, B., Beddington, R., Constantini, F., and Lacy, E. (ed.) *Manipulating the Mouse Embryo.* Cold Spring Harbor Laboratory Press, New York, 1994.

14. Smith, A. J., De Sousa, M. A., Kwabi-Addo, B., Heppell-Parton, A., Impey, H., and Rabbitts, P. (1994) A site-directed chromosomal translocation induced in embryonic stem cells by Cre-loxP recombination. *Nat. Genet.* **9,** 376–385.

15. Hunt, P., LeMaire, R., Embury, P., Sheean, L., and Mroz, K. (1995) Analysis of chromosome behavior in intact mammalian oocytes: monitoring the segregation of a univalent chromosome during female meiosis. *Hum. Mol. Genet.* **4,** 2007–2012.

16. Inoue, J., et al. (2001) Construction of 700 human/mouse A9 monochromosomal hybrids and analysis of imprinted genes on human chromosome 6. *J. Hum. Genet.* **46,** 137–145.

17. Shen, M. H., Yang, J., Loupart, M. L., Smith, A., and Brown, W. (1997) Human mini-chromosomes in mouse embryonal stem cells. *Hum. Mol. Genet.* **6,** 1375–2382.

18. Drost, J. B. and Lee, W. R. (1995) Biological basis of germline mutation: comparisons of spontaneous germline mutation rates among Drosophila, mouse, and human. *Environ. Mol. Mutagen.* **25,** 48–64.

12

Pronuclear Microinjection of Purified Artificial Chromosomes for Generation of Transgenic Mice

Pick-and-Inject Technique

Diane P. Monteith, Josephine D. Leung, Anita H. Borowski, Deborah O. Co, Tünde Praznovszky, Frank R. Jirik, Gyula Hadlaczky, and Carl F. Perez

1. Introduction

Mammalian artificial chromosomes (MACs) are appealing vectors for transgenesis and gene therapy applications. Their principal benefits include large DNA-carrying capacity and the ability to replicate in synchrony with the host genome. In addition, introducing genes into artificial chromosomes eliminates the disadvantages of integrating exogenous DNA into the host genome—variegated gene expression and insertional mutagenesis.

Several groups (*see* other chapters in this volume) have reported the production of mammalian artificial chromosomes by transfecting component DNA sequences that comprise centromeres and telom-

From: *Methods in Molecular Biology, Vol. 240:*
Mammalian Artificial Chromosomes: Methods and Protocols
Edited by: V. Sgaramella and S. Eridani © Humana Press Inc., Totowa, NJ

eres *(1–5)*; by telomere truncation of natural chromosomes *(6–15)*; by engineering natural chromosome fragments *(16)*; and using Epstein–Barr viral components to produce episomes *(17–22)*. We have reported the development and large-scale purification of satellite DNA-based artificial chromosomes—which we refer to as an Artificial Chromosome expression system (ACes) *(23–28)*.

ACes carry the necessary functional and structural DNA sequences possessed by natural chromosomes, including telomeres to ensure replication and protection of chromosome ends, centromeres for segregation during mitosis and meiosis, and replication origins distributed throughout their lengths to facilitate duplication of the genetic material during S-phase *(25,26,29)*. Currently more than 20 different ACes have been generated that carry genes encoding monoclonal antibodies, soluble receptors, therapeutic proteins, and green or red fluorescent proteins.

We have reported the generation of transgenic mice by pronuclear microinjection of ACes (30). Fluorescent *in situ* hybridization analyses of metaphase chromosomes from mitogen-activated peripheral blood lymphocytes from founder and progeny transgenic mice revealed that the ACes were maintained as discrete chromosomes and had not integrated into any host chromosomes. Studies conducted for up to 16 mo demonstrated stable ACes maintenance in peripheral blood lymphocytes. Currently, two lines of transgenic mice with prototype ACes have been produced. In the first line the prototype ACes has been passed through four generations in the germline (**Fig. 1A**, B-ACes). In the second line we have demonstrated tissue specific gene expression (**Fig. 1B**, A-ACes). Initial breeding data from both mouse lines indicates that ACes germline transmission is 50% through the females and about 4% for the males. Female germline transmission is consistent with Mendelian inheritance for a chromosome present at the single copy per cell level. The reason for the lower frequency transmission through the male is unclear, but we note that it is consistent with the range of male germline transmission frequencies observed in transgenic mice carrying univalent human chromosome fragments *(31–33)*, human small accessory chromosomes *(16)*, and human/murine chromosome chimeras *(15)*.

Previous groups have used microcell mediated chromosome transfer (MMCT) to generate MAC transgenic mice *(15,16,31–33)*, a procedure that is very inefficient and tedious. Artificial chromosomes were transferred via MMCT to murine embryonic stem (ES) cells then candidate drug-resistant ES cells were screened, analyzed, and transferred to murine blastocysts, which in turn were implanted in pseudopregnant females. ACes, in contrast, can be easily isolated—to purities exceeding 99%—by flow cytometry and microinjected into the pronuclei of fertilized oocytes.

The size of ACes (1 micron × 2 microns) presents unique disadvantages and advantages for the generation of transgenic mice. We invested a great amount of effort customizing microinjection pipets for optimum sharpness (bevel angle) and inner diameter size—wide enough to pass ACes—while still maintaining oocyte viability postinjection. ACes can be easily seen under differential interference optics (DIC) optics, during front loading (pick) into the microinjection pipet, and sliding down the pipet barrel into the pronucleus (inject).

Standard plasmid microinjection protocols *(34)* require the operator to observe the pronucleus to swell to almost double in size in to verify the introduction of the DNA solution. We initially used this endpoint for our pick and inject technique and found that oocyte viability postinjection was 29%. Now we terminate injection after we observe the ACes entering the pronucleus (without pronuclear swelling), attaining viabilities of 67%.

We currently produce founders at a 13% transgenesis frequency, which is comparable with the frequencies generated by the pronuclear injection of plasmid DNA. Procedures have been meticulously described for sorting ACes and chromosomes *(28,35,36)* and for maintaining a murine vivarium, harvesting murine oocytes, implanting injected oocytes, and analyzing progeny *(34,37)*. This chapter focuses on the procedures we pioneered to produce ACes transgenic mice.

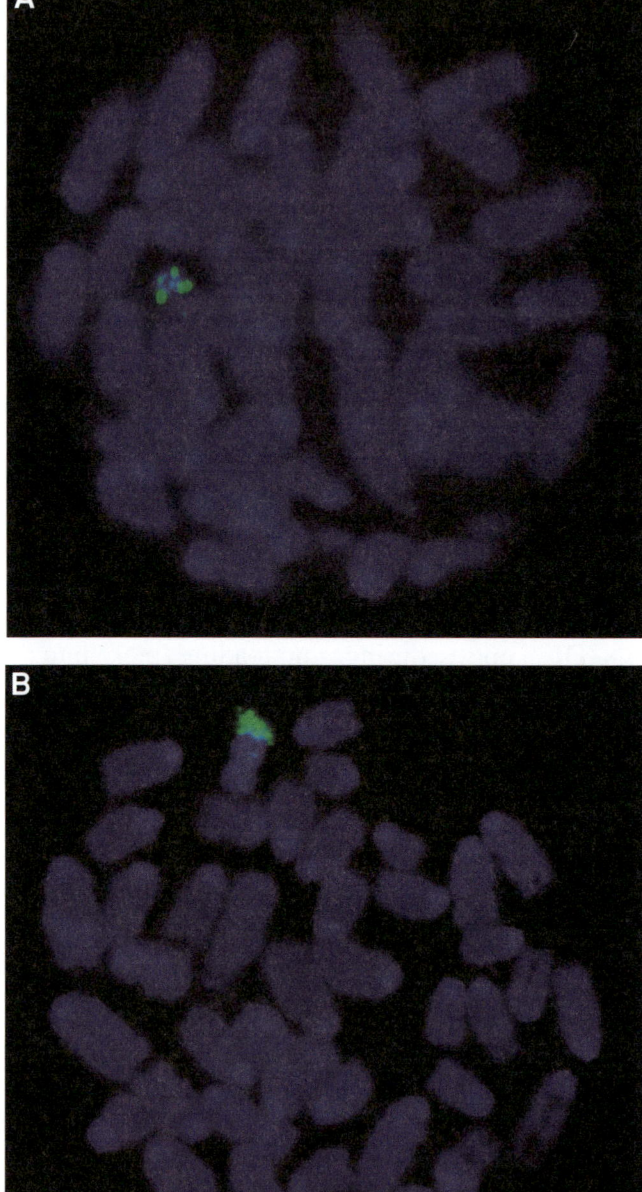

Fig. 1.

2. Materials

2.1. Quality Control of Microinjection Pipets

1. Customized microinjection pipets (2.3–3.2 μm outer diameter [OD], Humagen, Charlottesville, VA).
2. Plasticene (any office supply or hardware store).
3. Microscope slides.
4. Light microscope (×63 objective, ×10 eyepiece, and digital camera).
5. Compressed air and regulator.
6. Protractor.
7. Microscope eyepiece graticule (precalibrated).
8. 70% Ethanol (prepared from ultra pure water and pure grain alcohol).

2.2. Siliconization of Microinjection Pipets

1. Microinjection pipets (subjected to the quality control process outlined in **Subheading 3.1.**).
2. Siliconizing reagent (Dow Corning 1107 fluid).
3. Equilibrating syringe SAS11/2 manual air-driven (Research Instruments, Cornwall, UK).
4. Inverted light microscope.
5. Acetone, high-performance liquid chromatography grade (Fisher).
6. Trichloroethylene, 99.5% pure, spectrophotometric grade (Aldrich).

Fig. 1. Fluorescent *in situ* hybridization analyses of peripheral blood lymphocytes obtained from ACes transgenic mice. Blood (0.1 mL) was aseptically collected from transgenic mice by saphenous vein bleeding and cultured in a humidified atmosphere for 3 d at 37°C in RPMI supplemented 20% fetal bovine serum (with mitogens). Cultures were incubated for an additional 2 h in the presence of colcemid. Metaphase spreads from the cultured lymphocytes were hybridized to sequence-specific DNA probes and then counterstained with DAPI. (**A**) Metaphase spread of a B-ACes lymphocyte probed with hygromycin phosphotransferase (hph) gene sequences (green). Note that the probe detects four staining regions on this ACes-one on both arms of each of the two sister chromatids. (**B**) Spread of an A-ACes lymphocyte probed with a 40-kb cosmid encoding a therapeutic gene (green). The A-ACes contains 25 tandem copies of this cosmid for an approximate payload of 1 million bp.

7. Plasticene.
8. Ultra pure H_2O.
9. Drying oven.
10. Flask holder.
11. 100-mL beaker or equivalent glass recipient.

2.3. Preparation of Artificial Chromosomes and Transgenic Facility Requirements

1. ACes isolation facility. The equipment and procedures required for the isolation of ACes by bivariate flow cytometry have been described *(28)*. ACes are maintained in murine–viral–antibody-free LMTK⁻ cells or Chinese Hamster Ovary cells as verified by MAP testing (Institut de Armand Frappier, Canada).
2. Transgenic facility requirements and operation. The facility and procedures required for the generation of transgenic mice by microinjection of ACes are identical to those for the microinjection of DNA, have been described in detail *(34,37)*. We collect mouse zygotes from superovulated (C57BL/6 × CBA) F1 females approx 12 h after mating. All mice are viral–antibody free and are maintained in facilities that conform to specific pathogen-free standards.
3. Refrigerated swing-bucket centrifuge (Beckman, GS-60)

2.4. Microinjection of Fertilized One-Cell Embryos

1. 60-mm nontreated tissue culture dishes.
2. Depression slide (siliconized).
3. Mineral oil.
4. Disposable 3-cc syringes.
5. 0.2-μm filter (Millipore).
6. M16 culture media (Sigma).
7. M2 culture media (Sigma).
8. Injection pipets (subjected to the quality control process outlined in **Subheading 3.1.** and the siliconization process outline in **Subheading 3.2.**).
9. Holding pipets (straight, 65–95 μm OD, Humagen, Charlottesville, VA).
10. Capillary tube mouth pipetter (with filter).
11. Glass transfer pipet.

12. Micromanipulator (mechanical, Narishige, Tokyo).
13. Inverted microscope (Leica DM-IRB HC with DIC and phase contrast, ×5, ×20, and ×40 objectives, Leica Microsystems, Canada).
14. Vibration-free table, on which the microscope is mounted.
15. Equilibrating syringe SAS11/2 manual air-driven (Research Instruments, Cornwall, UK).
16. Dissecting microscope.
17. CO_2 tissue culture incubator (37°C, 5% CO_2).

3. Methods

3.1. Quality Control of Microinjection Pipets

1. In a clean and dust free environment, clean all supplies before use with 70% ethanol and remove any dust or lint by blowing with compressed air.
2. Lay the injection pipet horizontally on a microscope slide and secure it near the blunt end by a small piece of plasticene.
3. Observe the tip of the pipet under a ×630 magnification (do not use immersion oil). Make sure that the beveled edge is visible (refer to **Fig. 2**), otherwise rotate the pipet appropriately until the indicated edge is obtained.
4. Measure the size of the hypotenuse (h) of the pipet by aligning it against the graticule scale on the eyepiece. This value is expressed in graticule units.
5. Record image of the pipet. Print the image and measure the angle (A) subtended by the beveled edge of the pipet using a protractor (*see* **Note 1**). Calculate the OD, in μm, of the pipet as shown:

Sample analysis calculation: OD = h × sinA × C (*see* **Fig. 2**)
Where:

    OD = Outer diameter, in μm
    h = Hypotenuse in graticule units.

A = Angle subtended by the beveled edge of the pipet.
C = 1.57 μm/graticule-unit, conversion factor for the graticule scale at ×630 magnification.

Acceptable range of OD = 2.3 to 3.2 μm
Acceptable range of beveled angle = 25 to 40°

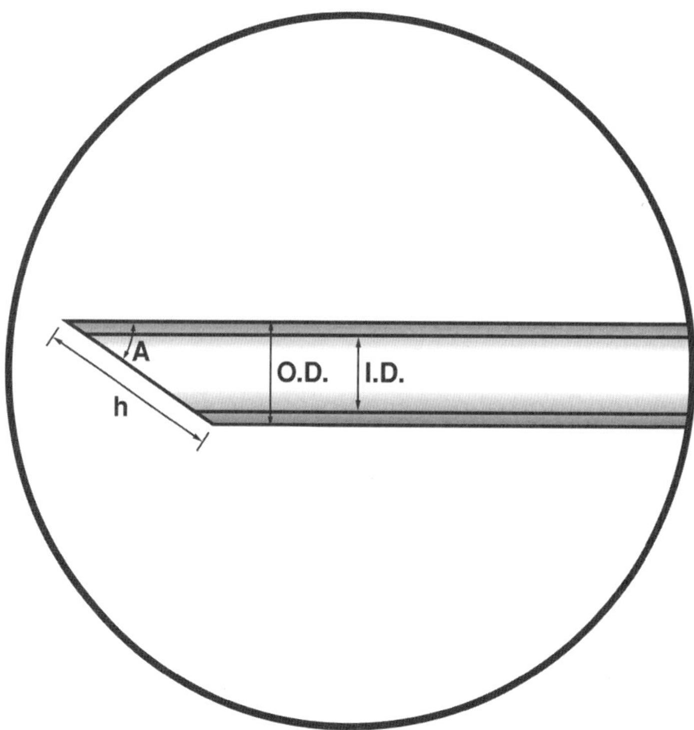

Fig. 2. Schematic for calculating OD of microinjection pipet. *See* text
(**Subheading 3.1.**) for discussion.

3.2. Siliconization of Microinjection Pipets

1. Using a fume hood, prepare solution by mixing 0.8 mL of Dow Corning 1107 fluid in 80 mL of trichloroethylene in a 100-mL beaker.
   Pinch a clump of plasticene around the mouth of the beaker.
2. Dip the beveled tip of the pipet into the siliconization solution 1–2 cm
   below the surface. Hold vertically in place by pressing the pipet into
   the clump of plasticene rimming the mouth of the beaker. Siliconization
   solution will rise into the pipet by capillary action (*see* **Note 2**).
3. Purge the siliconization solution from the pipet using the equilibrating syringe, monitoring under the light microscope at ×100 to ensure
   no solution remains in the pipet.
4. Dip the pipet in acetone for 10 s then purge the acetone from the
   pipet using the equilibrating syringe, monitoring under the light
   microscope at ×100 to ensure no acetone remains in the pipet.

5. Without removing the pipet from the equilibrating syringe, dip the pipet in water and draw water into the pipet by turning the knob counter-clockwise two full turns. Confirm by microscopy that water is drawn into the pipet.

6. Press the button on top of the knob to release the pressure inside the system then turn the knob clockwise to purge all liquid from the pipet.

7. Arrange the siliconized pipets on a flask holder by carefully sliding the pipets in between the coil and bake in the drying oven at 121°C for 1 h, then allow to cool to room temperature (*see* **Note 3**).

3.3. Preparation of Artificial Chromosomes

1. Place 1 million sorted ACes (*see* **Note 4**) into a 1.5-mL microfuge tube and concentrate by centrifugation in a swinging bucket rotor at 2500g for 15 min at 4°C (*see* **Note 5**).

2. Remove all but 20 μL of supernatant and store on ice. Concentration is routinely 1×10^4 ACes/μL.

3.4. Microinjection of Fertilized One-Cell Embryos

1. Filter-sterilize 2–3 mL of M2 media into a 60-mm tissue culture dish using a 0.2-μm filter and syringe.

2. Using a mouth pipet (with in-line filter), transfer 20 fertilized one-cell embryos from the lot provided for the session to the dish containing the filtered M2 media (*see* **Note 6**). Swirl embryos several times to wash.

3. Clean and wipe a siliconized depression slide with 70% ethanol and a lint-free wipe. Place an 8-μL droplet of M2 media onto the depression slide.

4. Flick the tube of concentrated ACes 10 to 20 times to resuspend the ACes and add 4 μL of the ACes suspension to the media droplet.

5. Overlay with mineral oil ensuring complete coverage of the droplet to reduce evaporation.

6. Transfer all washed embryos to the media/ACes droplet.

7. Carefully place the slide onto microscope stage.

8. Screw on the holding pipet to the left-hand instrument tube, then clip into position on the left-hand side of the micromanipulator.

9. Using first ×50, ×200, then ×400 magnification, center the holding pipet into the field of view by adjusting the fine controls of the micromanipulator.

10. Attach the microinjection pipet to the right-hand instrument tube and clip in place on the right-hand side of the micromanipulator. Center the field of view under ×50 magnification by adjusting the fine controls of the micromanipulator. Switch to ×200 magnification and turn the microinjection pipet so that the beveled edge appears to be in the down position. Switch to ×400 magnification while very gently orientating the microinjection pipet to the optimal position with the beveled edge down.

11. Further lower the pipet to same plane of focus as the holding pipet.

12. At ×200 magnification, turn the piston slowly in a clockwise direction to establish a controlled flow of media out of pipet (*see* **Note 7**).

13. Switching to ×400 magnification, slowly and carefully draw media into the microinjection pipet, always taking care not to build-up too much pressure in the system. Once control is achieved (*see* **Note 8**), the level of the media should be visible at such a distance up the microinjection pipet that it is within your field of view under ×400 magnification (*see* **Note 9**).

14. Pick up an embryo with the holding pipet and orientate the embryo so that a pronucleus can be clearly delineated and with no polar bodies obstructing the path of the pipet to the pronucleus (*see* **Note 10**).

15. Lower the embryo toward the bottom of the slide until it just makes contact—this step will support the embryo during the injection.

16. After the embryo is immobilized on the holding pipet, slowly draw up a single ACes into the injection pipet, then immediately redirect the pressure to move the ACes back down to the end of the microinjection pipet (*see* **Note 11**).

17. Establish control of the ACes by moving it slowly up and down the microinjection pipet with a steady intake and outflow of pressure. Then, lower the ACes as close as possible to the tip of the microinjection pipet.

18. Bring the tip of the microinjection pipet into the same plane of focus as the pronucleus of the immobilized embryo.

19. Carefully poke the needle through the zona pellucida, into the pronucleus. Eject the ACes into the pronucleus and then carefully withdraw the microinjection pipet back out of the embryo. Continue injecting all the eggs in the batch (*see* **Note 12**).

20. Transfer the intact embryos to a designated dish of M16 media and culture in a 5% CO_2 tissue culture incubator until time of implanta-

tion. Eggs can be transferred in the surrogate recipient on the same day as injection or they can be cultured overnight then transferred when most will have developed to the two-cell stage *(37)*.

3.5. Analyses of Transgenic Mice

1. Tail clippings are taken after weaning (*see* **Note 13**) following standard techniques *(34,37)*. DNA extracted from the tail tissue is subjected to PCR analysis using 3 sets of amplimers that are specific for ACes DNA sequences.
2. Blood (0.5–1.0 mL) is collected from transgenic mice (*see* **Note 14**) by saphenous vein bleeding *(38)* and the peripheral blood lymphocytes are cultured following standard procedures *(30)*.
3. Metaphase spreads are generated from peripheral blood lymphocytes *(39)* and analyzed for presence and integrity of ACes (*see* **Note 15**) by fluorescence *in situ* hybridization *(40)*.

4. Notes

1. The aspect ratio, 0.75, is the ratio of the inner diameter (ID) of the pipet compared to the outer diameter (OD) of the pipet. The aspect ratio remains constant for a micropipet—even when pulled on the microforge. During the OD calculation, we assume that the outer diameter is essentially constant for 10 μm from the tip of microinjection pipet.
2. Pipets may be left in the solution for up to 10 min so that batches (maximum of 4/batch) can be processed at the same time.
3. Store siliconized pipets in the original case for up to 2 mo at room temperature.
4. ACes are isolated by flow cytometry into sheath buffer at a concentration of 10^6 ACes/0.8 mL. We use presterilized, DNase-free, RNase-free, aerosol-resistant tips (20, 200, and 1000 μL; Fisher) when pipetting solutions containing ACes. ACes are gently transferred to avoid breakage.
5. Greater centrifugation speeds or use of a nonrefrigerated centrifuge will lead to ACes aggregation. Only single ACes should be picked and injected. In our experience, multiple ACes drawn up in the microinjection pipet obstruct the flow of fluid through the narrow channel at the tip of the pipet.

6. Embryos should be maintained in M2 media for no longer than 1 h. Therefore batches of 20 embryos can be managed at one time.

7. A volume of media is usually present within the microinjection pipet via capillary action. It is critical not to build up too much pressure in the system, which will introduce air bubbles into your droplet. If no media are present in needle, slowly and carefully draw up media by turning the piston in a counter-clockwise direction. Immediately stop turning the piston once media begins moving up the pipet. If no media are flowing into the pipet, the pipet may be improperly attached or have an obstruction and therefore the pipet should be reattached or replaced. A sudden uptake of suspension will create a plug of ACes at the opening of the microinjection pipet that may or may not be cleared (*see* **Note 9**).

8. Control is achieved once slow turns of the piston in both directions result in controlled movements of the ACes up and down the pipet. Minimize the amount of manipulations with the ACes in the micro-injection pipet. We have preliminary data that suggest that repeated handling of the ACes leads to breakage, which results in truncated ACes being microinjected and maintained in transgenic mice.

9. An ACes that is stuck in the microinjection pipet can sometimes be cleared by sudden but controlled uptakes of media that may dislodge it up into the pipet or by a forceful but controlled outflow of media that may expel it out of the opening. If an ACes aggregate forms that cannot be moved in either direction, you will have to replace your microinjection pipet in order to continue.

10. If an embryo is unfertilized or irregular, place it in a pile to the lower left of your uninjected embryos. The larger of the pronuclei (the male) is usually injected.

11. The ACes are floating freely in the M2 droplet. The critical skill of the technique is the slow approach of the microinjection pipet tip towards the desired ACes combined with the gradual turn of the piston to pick the ACes into the pipet, a skill that even an experienced injectionist needs to practice.

12. Place each intact embryo—the pronuclei should also be intact—in a pile at the upper right side of the region of the droplet.

13. Mice should be at least 3 wk old before tail clippings are taken.

14. Mice should be at least 5 wk old before they are subjected to saphenous tail bleeding. We routinely take blood samples every 2 mo.

15. At least 100 metaphase lymphocyte spreads should be scored.

Acknowledgments

The authors express their appreciation to Gary de Jong, Neil MacDonald, Michael Lindenbaum, Teresa McKernan, and Edmond Lee for their technical assistance and to Alexisann Maxwell, Ross Durland, and Helen Zeitler for helpful comments during manuscript preparation. The animal work was performed at both the Biomedical Research Centre and the Centre for Molecular Medicine and Therapeutics, University of British Columbia, Vancouver, British Columbia, Canada. This work is supported by Chromos Molecular Systems Inc., Burnaby, BC, Canada.

References

1. Harrington, J. J., Van Bokkelen, G., Mays, R. W., Gustashaw, K., and Willard, H. F. (1997) Formation of de novo centromeres and construction of first-generation human artificial microchromosomes. *Nat. Genet.* **15,** 345–355.
2. Warburton, P. E. and Cooke, H. J. (1997) Hamster chromosomes containing amplified human alpha-satellite DNA show delayed sister chromatid separation in the absence of de novo kinetochore formation. *Chromosoma* **106,** 145–159.
3. Ikeno, M., Grimes, B., Okazaki, T., Nakano, M., Saitoh, K., Hoshino, H., et al. (1998) Construction of YAC-based mammalian artificial chromosomes. *Nat. Biotech.* **16,** 431–439.
4. Henning, K. A., Novotny, E. A., Compton, S. T., Guan, X. Y., Liu, P. P., and Ashlock, M. A. (1999) Human artificial chromosomes generated by modification of a yeast artificial chromosome containing both human alpha satellite and single-copy sequences. *Proc. Natl. Acad. Sci. USA* **96,** 592–597.
5. Ebersole, T. A., Ross, A., Clark, E., McGill, N., Schindelhauer, D., Cooke, H., and Grimes, B. (2000) Mammalian artificial chromosome formation from circular alphoid input DNA does not require telomere repeats. *Hum. Mol. Genet.* **9,** 1623–1631.
6. Carine, K., Solus, J., Waltzer, E., Manch-Citron, J., Hamkalo, B., and Scheffler, I. (1986) Chinese hamster cells with a minichromosome containing the centromere region of human chromosome 1. *Somat. Cell Mol. Genet.* **12,** 479–491.

7. Farr, C., Fantes, J., Goodfellow, P., and Cooke, H. (1991) Functional reintroduction of human telomeres into mammalian cells. *Proc. Natl. Acad. Sci. USA* **88,** 7006–7010.

8. Farr, C. J., Stevanic, M., Thomson, E. J., Goodfellow, P. N., and Cooke, H. J. (1992) Telomere-associated chromosome fragmentation: applications in genome manipulation and analysis. *Nat. Genet.* **2,** 275–282.

9. Barnett, M. A., Buckle, V. J., Evans, E. P., Porter, A. C., Rout, D., Smith, A. G., and Brown, W. R. (1993) Telomere directed fragmentation of mammalian chromosomes. *Nucleic Acids Res.* **21,** 27–36.

10. Farr, C. J., Bayne, R. A., Kipling, D., Mills, W., Critcher, R., and Cooke, H. J. (1995) Generation of a human X-derived minichromosome using telomere-associated chromosome fragmentation. *EMBO J.* **14,** 5444–5454.

11. Heller, R., Brown, K. E., Burgtorf, C., and Brown, W. R. (1996) Minichromosomes derived from the human Y chromosome by telomere directed chromosome breakage. *Proc. Natl. Acad. Sci. USA* **93,** 7125–7130.

12. Shen, M. H., Yang, J., Loupart, M. L., Smith, A., and Brown W (1997) Human mini-chromosomes in mouse embryonal stem cells. *Hum. Mol. Genet.* **6,** 1375–1382.

13. Au, H. C., Mascarello, J. T., and Scheffler, I. E. (1999) Targeted integration of a dominant neo^R marker into a 2- to 30-Mb human minichromosome and transfer between cells. *Cytogenet. Cell Genet.* **86,** 194–203.

14. Mills, W., Critcher, R., Lee, C., and Farr, C. J. (1999) Generation of an approximately 2.4 Mb human X centromere-based minichromosome by targeted telomere-associated chromosome fragmentation in DT40. *Hum. Mol. Genet.* **8,** 751–761.

15. Shen, M. H., Mee, P. J., Nichols, J., Yang, J., Brook, F., Gardner, R. L., et al. (2000) A structurally defined mini-chromosome vector for the mouse germ line. *Curr. Biol.* **10,** 31–34.

16. Voet, T., Vermeesch, J., Carens, A., Durr, J., Labaere, C., Duhamel, H., et al. (2001) Efficient male and female germline transmission of a human chromosomal vector in mice. *Genome Res.* **11,** 124–136.

17. Krysan, P. J., Haase, S. B., and Calos, M. P. (1989) Isolation of human sequences that replicate autonomously in human cells. *Mol. Cell. Biol.* **9,** 1026–1033.

18. Sun, T. Q., Fernstermacher, D. A., and Vos, J. M. (1994) Human artificial episomal chromosomes for cloning large DNA fragments in human cells. *Nat. Genet.* **8,** 33–41.

19. Sun, T. Q., Livanos, E., and Vos, J. M. (1996) Engineering of a mini-herpesvirus as a general strategy to transduce up to 180 kb of functional self-replicating human mini-chromosomes. *Gene Ther.* **3,** 1081–1088.

20. Calos, M. P. (1996) The potential of extrachromosomal replicating vectors for gene therapy. *Trends Genet.* **12,** 463–466.

21. Simpson, K., McGuigan, A., and Huxley, C. (1996) Stable episomal maintenance of yeast artificial chromosomes in human cells. *Mol. Cell. Biol.* **16,** 5117–5126.

22. Kelleher, Z. T., Fu, H., Livanos, E., Wendelburg, B., Gulino, S., and Vos, J. M. (1998) Epstein-Barr-based episomal chromosomes shuttle 100 Kb of self-replicating circular human DNA in mouse cells. *Nat. Biotechnol.* **16,** 762–768.

23. Hadlaczky, G., Praznovszky, T., Cserpán, I., Keresö, J., Péterfy, M., Kelemen, I., et al. (1991) Centromere formation in mouse cells cotransformed with human DNA and a dominant marker gene. *Proc. Natl. Acad. Sci. USA* **88,** 8106–8110.

24. Praznovszky, T., Keresö, J., Tubak, V., Cserpán, I., Fátyol, K., and Hadlaczky, G. (1991) De novo chromosome formation in rodent cells. *Proc. Natl. Acad. Sci. USA* **88,** 11042–11046.

25. Keresö, J., Praznovszky, T., Cserpán, I., Fodor, K., Katona, R., Csonka, E., et al. (1996) De novo chromosome formations by large-scale amplification of the centromeric region of mouse chromosomes. *Chrom. Res.* **4,** 226–239.

26. Holló, G., Keresö, J., Praznovszky, T., Cserpán, I., Fodor, K., Katona, R., et al. (1996) Evidence for a megareplicon covering megabases of centromeric chromosome segments. *Chrom. Res.* **4,** 240–247.

27. Csonka, E., Cserpán, I., Fodor, K., Holló, G., Katona, R., Keresö, J., et al. (2000) Novel generation of human satellite DNA-based artificial chromosomes in mammalian cells. *J. Cell Sci.* **113,** 3207–3216.

28. de Jong, G., Telenius, A. H., Telenius, H., Perez, C. F., Drayer, J. I., and Hadlaczky, G. (1999) Mammalian artificial chromosome pilot production facility: large-scale isolation of functional satellite DNA-based artificial chromosomes. *Cytometry* **35,** 129–133.

29. Telenius, H., Szeles, A., Keresö, J., Csonka, E., Praznovszky, T., Imreh, S., et al. (1999) Stability of a functional murine satellite DNA-based artificial chromosome across mammalian species. *Chrom. Res.* **7,** 3–7.

30. Co, D. O., Borowski, A. H., Leung, J. D., van der Kaa, J., Hengst, S., Platenburg, G. J., et al. (2000) Generation of transgenic mice and

germline transmission of a mammalian artificial chromosome introduced into embryos by pronuclear microinjection. *Chrom. Res.* **8,** 183–191.

31. Tomizuka, K., Yoshida, H., Uejima, H., Kugoh, H., Sato, K., Ohguma, A., et al. (1997) Functional expression and germline transmission of a human chromosome fragment in chimaeric mice. *Nat. Genet.* **16,** 133–143.

32. Hernandez, D., Mee, P. J., Martin, J. E., Tybulewicz, V. L., and Fisher, E. M. (1999) Transchromosomal mouse embryonic stem cell lines and chimeric mice that contain freely segregating segments of human chromosome 21. *Hum. Mol. Genet.* **8,** 923–933.

33. Tomizuka, K., Shinohara, T., Yoshida, H., Uejima, H., Ohguma, A., Tanaka, S., et al. (2000) Double trans-chromosomic mice: maintenance of two individual human chromosome fragments containing Ig heavy and {k} _loci and expression of fully human antibodies. *Proc. Natl. Acad. Sci. USA* **97,** 722–727.

34. Murphy, D. and Carter, D. A. (eds.) Transgenesis Techniques. Humana, Totowa, NJ, 1993.

35. Carter, N. P. Bivariate chromosome analysis using a commercial flow cytometer, in *Chromosome Analysis Protocols* (Gosden, J. R., ed.), Humana, Totowa, NJ, 1994, pp. 187–204.

36. Fantes, J. A., Green, D. K., and Sharkey, A. Chromosome sorting by flow cytometry, in *Chromosome Analysis Protocols* (Gosden, J. R., ed.), Humana, Totowa, NJ, 1994, pp. 205–219.

37. Hogan, B., Beddington, R., Costantini, F., and Lacy, E. (eds.) *Manipulating the Mouse Embryo: A Laboratory Manual.* Cold Spring Harbor Press, Cold Spring Harbor, NY, 1994.

38. Hem, A., Smith, A. J., and Solberg, P. (1998) Saphenous vein puncture for blood sampling of the mouse, rat, hamster, gerbil, guinea pig, ferret, and mink. *Lab. Anim.* **32,** 364–368.

39. Dracopoli, N. C., Haines, J. L., Korf, B. R., Moir, D. T., Morton, C. C., Seidman, C. E., et al. Clinical cytogenetics, in *Current Protocols in Human Genetics.* John Wiley and Sons, NY, 1994.

40. Pinkel, D., Straume, T., and Gray, J. W. (1986) Cytogenetic analysis using quantitative, high-sensitivity, fluorescence hybridization. *Proc. Natl. Acad. Sci. USA* **83,** 2937–2938.

13

Modification of Human Bacterial Artificial Chromosome Clones for Functional Studies and Therapeutic Applications

Michael R. Orford, Duangporn Jamsai,
Samuel McLenachan, and Panayiotis A. Ioannou

1. Introduction

With the completion of the sequencing phase of the Human Genome Project and the availability of fully sequenced clones containing intact functional loci from the various high-quality human P1- and F-plasmid derived/bacterial artificial chromosome (PAC/ BAC) genomic libraries *(1,2)*, it is becoming more convenient to study the expression and regulation of genes in their native genomic environment and to use such knowledge for therapeutic applications.

However, a major limitation in such studies has been the difficulty of introducing specific modifications in these large PAC/BAC clones that, for obvious reasons, cannot be efficiently manipulated by standard cloning procedures. The development of various techniques using homologous recombination in *Escherichia coli* has been invaluable in overcoming these limitations (**refs. *3–16*, for a recent review, *see* **ref.** *17*) and has allowed high precision modifications to be performed irrespective of the availability of restriction

From: *Methods in Molecular Biology, Vol. 240:*
Mammalian Artificial Chromosomes: Methods and Protocols
Edited by: V. Sgaramella and S. Eridani © Humana Press Inc., Totowa, NJ

sites. Such modifications include the insertion or deletion of sequences from a few base pairs to at least 70 kb in length, the introduction of reporter genes into any desired location in intact loci, the introduction of physiologically relevant or disease causing mutations and the direct subcloning of BAC fragments into smaller expression vectors. However, not all methods are readily applicable or convenient for the modification of BACs because they are either not directly applicable in the *E. coli* DH10B strain in which most PAC/BAC libraries are maintained or they require the construction of shuttle plasmids. The *E. coli* DH10B strain has a number of features, including a mutation in the *recA* gene, that make it the strain of choice for the stable maintenance of large genomic inserts in PAC/BAC clones. However, this also means that PAC/BAC clones cannot be modified by homologous recombination in DH10B cells without restoring recombination proficiency, thus potentially risking intramolecular rearrangements and deletions. Restoration of the RecA activity in a temperature-sensitive shuttle plasmid was originally used to modify BACs in DH10B cells *(3)*. The subsequent development of the RecE/RecT (ET) cloning system for the inducible restoration of recombination proficiency in *E. coli (5)* has led to a number of more convenient and efficient methods for the modification of PAC/BAC clones.

The GET Recombination system *(6,10,11)* that is the subject of this chapter has been specifically designed to facilitate modification of PAC/BAC clones in DH10B cells and allows the introduction of a variety of modifications without the need to create special shuttling vectors.

2. Materials

1. *E. coli* strain DH10B (Life Technologies–Invitrogen).
2. Luria/Bertani (LB) media.
3. Antibiotic stock solutions.
4. UV/Visible Spectrophotometer.
5. Sterile 10% (w/v) L-arabinose solution.
6. Bench-top refrigerated centrifuge.

7. Refrigerated Eppendorf microcentrifuge.
8. Sterile 10% (v/v) glycerol solution.
9. Electroporation apparatus, for example, Bio-Rad Gene Pulser II.
10. 0.1-cm gap electroporation cuvets.
11. SOC medium.
12. Plasmid pGETrec.
13. Microbiological shakers and plate incubators.
14. Plasmids pUC19 and pZeoSV2.
15. Polymerase chain reaction (PCR) thermocycler.
16. Targeting and PCR screening primers.
17. Restriction enzyme *Dpn*I.
18. Agarose and pulse field gel electrophoresis apparatus.
19. Qiagen Gel Extraction Kit.
20. 3 *M* sodium acetate, pH 5.2.
21. Absolute and 70% (v/v) ethanol.
22. TE buffer.
23. Sterile water.

3. Methods

The methods section described below will outline the various stages involved in the generation of targeted modifications to BACs using the GET Recombination system. One example experiment will additionally be described to illustrate the experimental process involved, outlining the replacement of the ε-globin gene held in a 200-kb BAC clone with an EGFP-Kanamycin/Neomycin reporter/ selection cassette. The methods chapter will thus be divided into the following sections: (1) an overview of the GET Recombination system; (2) general electroporation protocols; (3) GET Recombination protocols; (4) the replacement of the ε-globin gene with an EGFP-Kanamycin/Neomycin reporter/selection cassette in clone pEBAC/ 148β; and (5) time considerations.

3.1. The GET Recombination System

The RecE/RecT and the bacteriophage λ Red systems promote homologous recombination between linear DNA fragments with as little as 30 bp of flanking sequence homology and circular plasmid molecules or the host chromosome *(4,5)*. We initially found that the ET cloning system *(5)* could not be used to modify BACs in the *E. coli* DH10B host cell strain because of the rapid degradation of linear DNA fragments by the highly active RecBCD nuclease. Moving BACs from the DH10B strain to other recombination proficient strains also proved impractical due to methylation restriction limitations. We therefore developed further the ET cloning system, to allow the modification of PACs/BACs directly in DH10B cells by the inducible inhibition of the RecBCD nuclease *(6)*.

3.1.1. The pGETrec Plasmid

Plasmid pGETrec (for availability, *see* **Note 1**) was derived from plasmid pBAD24-tRecET of the ET cloning system *(5)* by the cloning of the bacteriophage λ *gam* gene downstream of the *recT* gene in a polycistronic operon, thereby placing the expression of the *gam* gene under the same L-arabinose inducible promoter as the RecE and *recT* genes (**Fig. 1, ref. *6***).

The *araBAD* promoter system allows rapid and homogeneous expression of proteins to high levels in *E. coli* cultures at saturating concentrations of L-arabinose *(18,19)*. It displays a fast on-off behavior, being switched on by L-arabinose and switched off by glucose. Rapid accumulation of RecE/RecT proteins was seen in DH10B cells after induction with 0.2% w/v L-arabinose (**Fig. 2**), with the RecE/RecT protein band becoming the most intense band within 30–40 min of induction. Prolonged induction of DH10B (BAC, pGETrec) cells with L-arabinose was found to lead to a slowing of growth rate and cell death, while induction for only 10 min leads to a 10-fold reduction in the number of recombinant clones. Thus induction with L-arabinose is normally carried out for 40 min. The synthesis of RecE/RecT proteins is rapidly switched off after electroporation by dilution of cells in SOC because of the presence of glucose, but our studies indicate that the preformed RecE/RecT proteins can persist for at least 3 h at 37°C (data not shown).

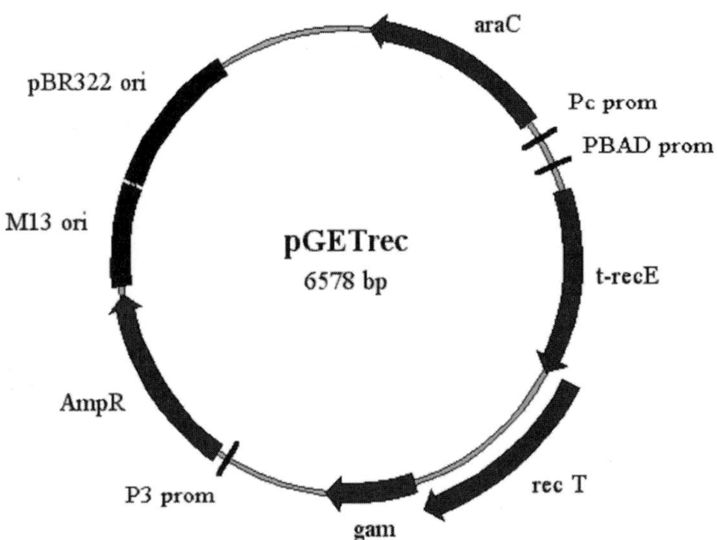

Fig. 1. Map showing the key features of the pGETrec plasmid. The *t-recE* and *recT* gene products provide inducible recombination proficiency to DH10B cells, whereas the *gam* gene product transiently inhibits RecBCD nuclease activity, thereby protecting incoming linear double- stranded PCR fragments from degradation.

The Gam protein is too small to resolve satisfactorily under the conditions used for the electrophoresis of the RecE/RecT proteins. However, isolation of PCR product at various time points after electroporation into DH10B (pGETrec) cells showed rapid elimination of linear DNA in uninduced cells, whereas induction of cells with 0.2% L-arabinose resulted in a significant protection of linear DNA from degradation, with some intact PCR fragment still detectable one hour after electroporation (data not shown). Thus, our results indicate that induction of the *recE* and *recT* genes from the pGETrec plasmid induces a recombinogenic state that may last for up to several hours, whereas specific recombination products may only be obtained after electroporation by slowing down the rapid degradation of the linear DNA by the RecBCD nuclease, through the induction of the *gam* gene.

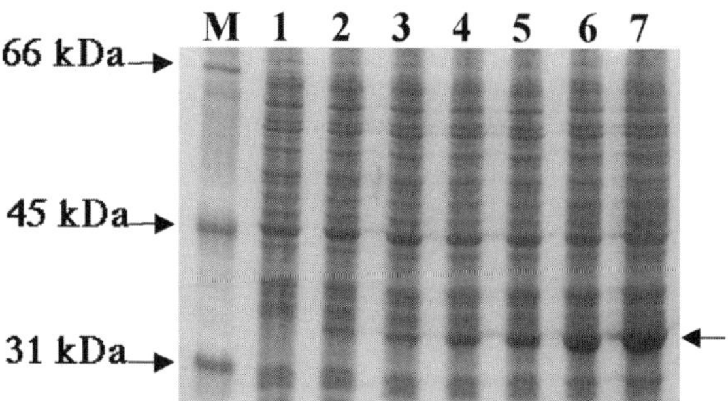

Fig. 2. Induction of RecE/RecT proteins from the pGETrec plasmid in DH10B cells. DH10B (pGETrec) cells were induced with 0.2% L-arabinose and 1.5-mL aliquots taken at 15, 20, 40, 60, 90, and 240 min after induction. The medium was removed, and the cells were lysed in 50 μL of sample loading buffer (250 m*M* Tris, pH 6.8, 2% sodium dodecyl sulfate; 10% glycerol; 20 mM DTT). A 5-μL aliquot was analyzed by discontinuous sodium dodecyl sulfate polyacrylamide gel electrophoresis by the method of Laemmli *(20)* using a Protean II minigel apparatus (Bio-Rad) and Coomassie Blue staining. The RecE and RecT proteins are expected to be very similar in size (about 33 kDa, indicated by the arrow) and are not resolved on electrophoresis of whole cell extracts. Lane M: Molecular weight marker; lane 1, noninduced cells; lanes 2–7, cells at 15, 20, 40, 60, 90, and 240 min after arabinose induction.

3.1.2. Design of Targeting Cassettes

Targeting cassettes for homologous recombination are conveniently prepared by PCR using customized targeting primers. Each primer is designed with two functional elements serving distinct roles: Firstly, a 20- to 30-bp sequence at the 3' end is required to prime amplification of the cassette from the PCR template; second, a 40- to 50-bp sequence at the 5' end of each primer is homologous to a unique target sequence on the BAC clone. The two 5' sequences together define the region of the BAC clone that will be targeted during homologous recombination (**Fig. 3**). It is particularly important to avoid repetitive elements in the sequences of the targeting oligonucleotides because perfect regions of homology less than 50 bp in the BAC sequence can readily be recognized as targets for recombina-

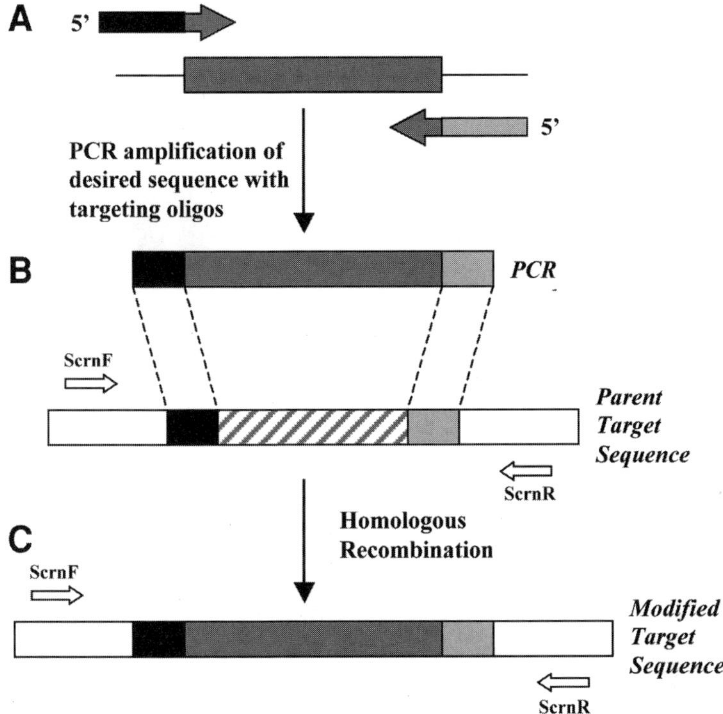

Fig. 3. Overview of the GET Recombination system. (**A**) Oligonucle-otide primers are designed to prime amplification of the sequence to be inserted into the BAC, and target its insertion to the desired region of the BAC clone. (**B**) The targeting cassette pairs with its complementary sequences on the BAC clone and undergoes recombination to yield the final modification. (**C**) Screening PCR primers (ScrnF and ScrnR) amplify across the recombination junction for the purpose of detection of modified clones because of different size products produced in the modified and parent molecules.

tion and reduce the overall efficiency of the targeting event *(5,10)*. Longer regions of homology yield a higher frequency of recombi-nant clones *(5,6)*, but the use of such cassettes is not convenient because it requires multiple steps for their synthesis. The availabil-ity of large numbers of fully sequenced PAC and BAC clones from the Human Genome Project greatly facilitates the design of targeting oligonucleotides for the modification of such clones.

3.1.3. Outline of Experimental Procedures

The modification of PAC/BAC clones requires first the electroporation of the pGETrec plasmid into the same DH10B cells as the PAC/BAC clone. The isolation and growth of such cells is conductedwith continuous antibiotic selection for both the pGETrec and BAC plasmids, whereas induction with L-arabinose is conducted just before harvesting and preparation of the cells for electroporation. The targeting cassette is then electroporated into the DH10B (BAC, pGETrec) cells. Recombination between the targeting cassette and the BAC clone (**Fig. 3**) allows the generation of recombinant clones. Selection of recombinant clones is usually facilitated by the inclusion of an antibiotic gene in the targeting cassette that confers novel antibiotic resistance to the recombinant clones. The introduction of point mutations and other fine modifications without leaving behind any operational sequences is conducted in two separate stages *(11)* with the use of a counterselection marker to kill cells carrying nonrecombinant clones.

3.2. General Electroporation Protocols

3.2.1. Preparation of Electrocompetent E. coli DH10B Cells

1. Pick a single colony of DH10B cells from a freshly streaked plate and inoculate a 2-mL overnight starter culture (LB plus appropriate antibiotics; *see* **Note 2**).
2. The next day, inoculate 20 mL of media with 0.2 mL of the overnight culture and grow at 37°C until the culture reaches an OD_{600} of 0.5. DH10B (BAC, pGETrec) cells being prepared for GET Recombination should be grown at 37°C until they reach an OD_{600} of 0.4, after which the media should be supplemented with L-arabinose to a final concentration of 0.2% (w/v) by adding 0.4 mL of a 10% (w/v) filter-sterilized stock solution, followed by a further shaking for 40 min at room temperature.
3. Harvest the cells by centrifugation at 5000 rpm for 5 min, resuspend the pellet in 3 mL of ice cold 10% (v/v) glycerol, and transfer to two prechilled microcentrifuge tubes.

4. Centrifuge the cells at 0°C for 1 min at 13,000 rpm in a refrigerated microcentrifuge and pipet off the supernatant.
5. Resuspend each of the pellets in 1.5 mL ice cold 10% (v/v) glycerol and repeat the centrifugation step.
6. Repeat the washing cycles a further three times each time with 1.5 mL ice cold 10% (v/v) glycerol and resuspend the pellets in a final volume of 40 µL ice cold 10% (v/v) glycerol.
7. The cells (40 µL) can be used directly for electroporations or snap frozen in an ethanol/dry ice bath and stored at –70°C until needed (*see* **Note 3**).
8. Larger batches of cells can be prepared using the above protocol by scaling up the volumes as necessary and storing the aliquots at –70°C after snap freezing in a CO_2/ethanol bath.

3.2.2. Electroporations

1. Preferentially use freshly prepared cells or thaw electrocompetent cells on ice if using frozen cells prepared previously (*see* **Note 3**).
2. Add 1–2 µL precooled DNA in 0.5X TE buffer containing 0.5–2.0 µg PCR product for GET Recombination, or 100–200 pg of plasmid DNA for transformations to the cells and mix by tapping (*see* **Note 4**).
3. Transfer the cells/DNA mixture with a cooled pipet tip to a pre-chilled electroporation cuvet (0.1-cm gap) taking care to minimize the introduction of air bubbles into the chamber of the cuvet.
4. Electroporate the cells using 1.8 kV, 200 Ω, and 25 µF, which will result in a time constant of 4–5 ms.
5. Add 1 mL SOC medium to the cuvet, transfer the cells to a 15 mL tube and shake at 37°C for 1 h.
6. Plate the cells onto LB agar plates containing appropriate antibiotics and allow colonies to form over a 16- to 24-h period.

3.3. GET Recombination Protocols

The following sections outline the general procedures used in the GET Recombination system to introduce targeted modifications to BAC clones. **Subheading 3.4.** will refer to a specific example of use of the system.

3.3.1. Transformations

1. Pick a single colony of DH10B (BAC) cells and make them electrocompetent, as described in **Subheading 3.2.1.**
2. Add 1 μL of plasmid pGETrec (approx 100 pg) to the cells, electroporate and plate on LB agar plates containing antibiotic selection for the BAC clone and ampicillin for pGETrec selection (**Subheading 3.2.2.**).
3. Pick a colony of the DH10B cells, now containing both the BAC and pGETrec plasmids and grow an overnight starter culture at 37°C, including antibiotic selection for both the BAC and the pGETrec plasmids (*see* **Note 5**).
4. Use the overnight culture to inoculate a 20-mL culture of LB containing the same antibiotics at the same concentrations, grow the cells at 37°C, and prepare them for GET Recombination, including the additional 40-min induction step with L-arabinose as described in **Subheading 3.2.1.**

3.3.2. Electroporation Efficiency

The transformation efficiency of electrocompetent cells should be carefully evaluated. Commercially available DH10B cells (grown without antibiotics) should yield transformation efficiencies in the order 10^{10} cfu/μg DNA. This is rarely achievable with home made cells, whereas the preparation of electrocompetent DH10B (BAC, pGETrec) cells in the presence of two antibiotics has a further detrimental effect. Home made DH10B (BAC, pGETrec) cells, grown on ampicillin and chloramphenicol, should generally yield an efficiency in excess of 10^8 cfu/μg DNA.

1. To an aliquot of electrocompetent DH10B (BAC, pGETrec) cells, add a standard known amount of a commercially available plasmid, for example, 10 pg of pUC19 for cells not carrying the pGETrec plasmid or 10 ng of pZeoSV2 for cells that contain pGETrec.
2. Electroporate the cells as described in **Subheading 3.2.2.** and plate serial dilutions of the SOC incubations on LB agar plates containing appropriate antibiotics (ampicillin for pUC19 or zeocin for pZeoSV2 on low-salt LB [pH 7.5] plates).
3. Count the number of colonies and estimate the number of colony forming units (cfu/μg DNA electroporated).

3.3.3. Retention of Plasmids

We have observed a variable tendency for DH10B (BAC, pGETrec) cells to lose the BAC or the pGETrec plasmid during preparation of electrocompetent cells, thus affecting adversely the overall efficiency of the system. Empirical observations suggest that this tendency can be affected by the conditions of culture of the cells, including temperature of incubation and antibiotic concentration. The use of freshly streaked and well-growing colonies and the avoidance of overgrowing cells on plates and in culture should minimize this problem. Growth of cultures for electrocompetent cells at 30°C and increased antibiotic concentration may also improve retention in some circumstances. If low retention of both plasmids is suspected, careful evaluation of the cells should be performed as follows:

1. Dilute an aliquot of induced DH10B cells prepared for GET Recombination and plate on LB agar plates containing either ampicillin, the antibiotic used to select for the BAC clone (usually chloramphenicol), or no antibiotic.
2. After overnight growth, count the number of colonies on the plates and calculate the proportion that are resistant to each antibiotic. If the proportion of the total clones that are resistant to each antibiotic is less than 90%, the efficiency of the system can be significantly reduced and the cells may need to be prepared again.

3.3.4. Preparation of Targeting Cassettes for Electroporation

1. Set up 16 PCR reactions (25 μL) using the targeting oligonucleotide primers for the amplification of the sequence to be recombined into the BAC.
2. Run the PCR for 30 cycles (94°C, 30 s; 60°C, 30 s; 72°C, 1 min/kb length of PCR product; *see* **Note 6**).
3. Pool the reactions and digest for 1 h at 37°C with *Dpn*I. This will digest the methylated plasmid template thereby reducing plasmid carryover into the electroporations and the appearance of false positive clones (*see* **Note 7**).
4. Purify the PCR product using the Qiagen PCR purification kit (optionally the PCR product may be gel purified).

5. Precipitate the purified fragment by adding 0.1 vol of 3 *M* sodium acetate, pH 5.2, and 2 vol of absolute ethanol and leave on ice for 30 min.
6. Centrifuge at 4°C for 30 min at 13,000 rpm, decant the supernatant, and wash the pellet with 0.5 mL of 70% (v/v) ethanol.
7. Remove the supernatant, air-dry the pellet at room temperature for 5 min and redissolve in 5 μL of 0.5X TE buffer.
8. Quantitate the amount of PCR product by resolving 1 μL on an agarose gel and compare the intensity of the band to that of the standards of known concentrations.
9. Use 1–2 μL PCR product containing 0.5–2 μg per 40 μL of cells for electroporation, as described in **Subheading 3.2.2.**

3.3.5. Screening for Recombinant Clones

Initial screening of colonies can be performed by colony PCR as detailed below. However, care should be taken when picking the colonies to avoid contamination from the lawn of plated cells present on the surface of the plate that carry unmodified BAC DNA and may thus constitute a source of false-negative PCR signal. Alternatively, individual clones can be grown overnight and a 1-μL aliquot of the culture used directly for PCR. Screening primers should be designed to anneal outside of the region modified by GET Recombination to produce distinctly different PCR products in modified and unmodified clones.

1. Pick colonies from the plates after GET Recombination and place into 40 μL of sterile water (or LB medium), in individual wells of a 96-well plate.
2. Disperse the colonies by agitation, and use 1–5 μL of the solution as template for PCR in 25-μL PCR screening reactions.
3. Any identified positive clones can be used to inoculate overnight cultures in LB containing appropriate antibiotics for further analysis.

3.3.6. Postmodification Analysis

BAC clones modified by the GET Recombination system should be carefully analyzed to demonstrate: (a) that the desired modification has been performed and (b) that the BAC clones have not undergone any unwanted rearrangements during the process.

1. Remove plasmid pGETrec from the modified clone by preparing miniprep DNA from an overnight culture grown in the absence of ampicillin, electroporate 1 µL into fresh DH10B cells and plate the cells on LB agar plates, selecting only for the BAC clone (**Subheading 3.2.1.**).

2. Pick a single colony, prepare a miniculture and confirm loss of the pGETrec plasmid by virtue of ampicillin sensitivity. Alternatively, clones can be streaked out on chloramphenicol plates and transferred to ampicillin plates, to identify colonies that spontaneously loose the pGETrec plasmid.

3. From 2-mL overnight cultures of DH10B cells containing either the parent or the modified BAC clone, make miniprep DNA and perform a series of restriction enzyme digests by standard protocols, choosing two to three different enzymes that cut the BAC into small-, medium-, and large-sized fragments and analyze by agarose and pulse field gel electrophoresis as necessary.

4. Perform southern blot analysis with ^{32}P-labeled probes either consisting of the region of the BAC deleted, to demonstrate its loss, or the PCR product inserted, to demonstrate its integration into the BAC on the correct sized restriction enzyme fragment.

5. Sequence junction fragments.

3.4. Replacement of the ε-Globin Gene in the β-Globin Locus With an EGFP-Kan/Neo Reporter/Selection Cassette

3.4.1. Background

In this example, a modification to the genomic insert of pEBAC/ 148β *(6,10)*, a 200-kb second-generation BAC clone is described. The coding sequence of the ε-globin gene was precisely deleted between the start and stop codons and replaced with the EGFP-Kan/ Neo cassette such that the start codon of the EGFP gene was in direct replacement of the start codon of the ε-globin gene. In this way, EGFP expression is driven directly from the native ε-globin promoter, thereby providing an accurate reporter assay system for

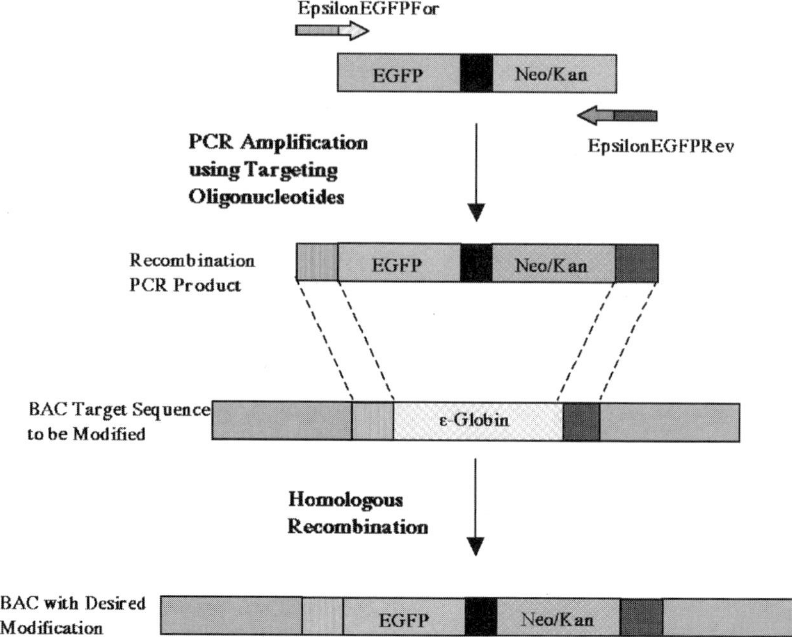

Fig. 4. Replacement of the ε-globin gene in pEBAC/148β with an EGFP-Kan/Neo cassette. The EGFP-Kan/Neo cassette from plasmid pEGFP-N22 was targeted by GET Recombination between the start and stop codons of the ε-globin gene in pEBAC/148β containing a 185-kb genomic fragment with the intact β-globin locus. Recombinant clones were identified by screening with the external primers EpsilonScrnF and EpsilonScrnR.

ε-globin expression in its natural genomic environment (**Fig. 4**). This construct complements a previously described construct *(10)* in which the insertion of the EGFP-Kan/Neo cassette was accompanied by a 44-kb deletion of all the sequences between the start codon of the ε-globin gene and the stop codon of the β-globin gene.

3.4.2. Preparation of EGFP-Kan/Neo Targeting Cassette

The EGFP-Kan/Neo cassette (2861 bp) was prepared for GET Recombination by PCR amplification from plasmid pEGFP-N22 using primers EpsilonEGFPFor (5' TCTGCTTCCGACACAGCT GCAATCACTAGCAAGCTCTCAGGCCTGGCATC atggtgagcaagggcgaggagc-3') and EpsilonEGFPRev (5'-CAGAA GGAGGGGTGTCAGGGTCACAGGAAGACCTGC AAACTGGAAGAGAACccagagtcccgctcagaag-3'). Upper case characters (50 bp) relate to the homology targeting arms flanking the ε-globin genomic sequence, and lower case characters those used to prime amplification of the EGFP-Kan/Neo cassette. Plasmid pEGFP-N22 was derived from plasmid pEGFP-N2 (Clontech) by deletion of the multicloning site and removal of the *Not*I site located downstream of the EGFP gene *(10)*. PCR reactions were performed in 25-μL volumes for 30 cycles (94°C, 30 s; 60°C, 30 s; 72°C, 2 min) with the Expand High Fidelity PCR System (Boehringer Mannheim). The resulting PCR product was *Dpn*I treated, gel purified, and precipitated as described in **Subheading 3.3.3.**

3.4.3. GET Recombination

DH10B (pEBAC/148β) cells were electroporated with pGETrec plasmid and prepared for GET Recombination as described in **Subheading 3.3.1.** The purified PCR product was electroporated and colonies allowed to form over an 18-h period (**Subheading 3.2.2.**). A number of kanamycin-resistant colonies that grew were picked and screened by PCR using primers EpsilonScrnF (5'-CACAGCTGCAATCACTAGCA-3') and EpsilonScrnR (5'-CCATCATATCA TCCTCCTTGG-3'). These primers yield a PCR product of 1766 bp from the normal ε-globin gene, while insertion of the EGFP-Kan/Neo cassette should yield a PCR product of 3106 bp (**Fig. 5**).

3.4.4. Postmodification Analysis

Miniprep DNA from two individual recombinant clones was prepared, re-electroporated into fresh DH10B cells and plated on LB agar plates containing chloramphenicol to segregate the pGETrec plasmid. Single colonies were subsequently picked, checked for loss

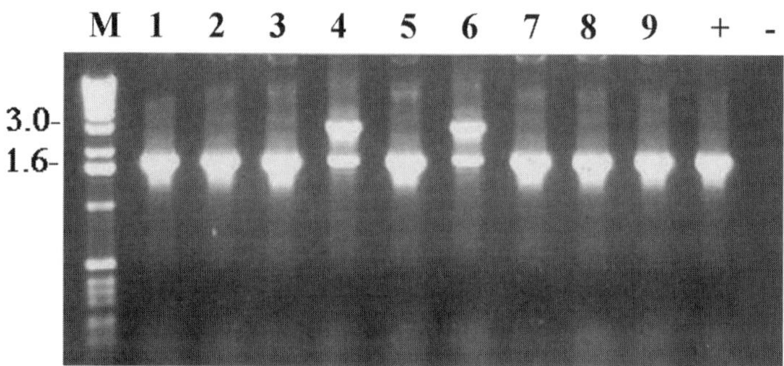

Fig. 5. PCR screening for recombinant clones. Nine individual colonies were screened by PCR using primers EpsilonScrnF and EpsilonScrnR to yield products of 1766 bp for unmodified clones, or 3106 bp for clones which had undergone the desired modification. Lanes 4 and 6 clearly identified two positive clones that had undergone the desired modification. Lanes + and – represent controls (pEBAC/148β, unmodified control and water, negative control, respectively). Note that the positive clones additionally amplify a small 1766-bp PCR product as an artefact of colony PCR (**Subheading 3.3.5.**).

of pGETrec by ampicillin sensitivity, and grown in overnight mini cultures. Miniprep DNA was prepared for restriction enzyme digest mapping experiments, to check for unwanted rearrangements at various levels of resolution (**Fig. 6**).

Digestion of pEBAC/148β DNA with *Not*I releases the 185-kb genomic insert from the 17-kb vector backbone when resolved on a 1% PFGE gel (**Fig. 6A**). Replacement of the ε-globin gene with the EGFP-Neo/Kan cassette is expected to give rise to an increase in size of 1340 bp, which is not resolved under these conditions in clones positive for this type of replacement. Digestion of unmodified pEBAC/148β using *Xho*I produces six fragments (**Fig. 6B**, lane 1). *Xho*I cuts only twice in the known sequence of the β-globin locus, releasing a 4.9-kb fragment between the two γ-globin genes. The uppermost 75-kb band contains most of the β-globin locus from the ^{A}γ-globin gene to the 3' end of the locus, whereas the 5' end of the locus, including the LCR and ε-globin gene, are contained in the upper band at about 45 kb (data not shown). The insertion of the EGFP-Neo/Kan cassette in place of the ε-globin gene should result

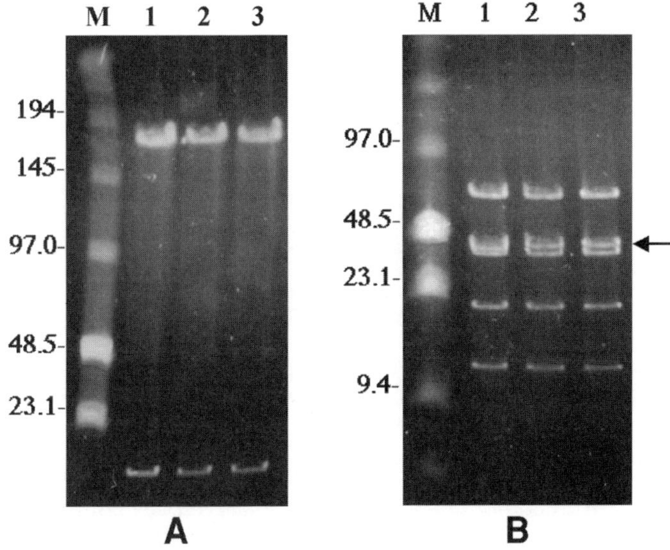

Fig. 6. Restriction enzyme mapping of modified clones at low and medium resolution. Two positive clones identified by PCR screening were digested with either *Not*I to release the entire 185-kb genomic insert from the 16-kb vector fragment (**A**) or digested into medium sized fragments with *Xho*I (**B**). No change is detectable after *Not*I digestion, whereas a small but clearly observable change in the size of the upper 45-kb band, carrying the ε-globin gene, is seen in the recombinant clones (arrow, lanes 2, 3) after *Xho*I digestion. Lane 1, Parent clone, pEBAC/ 148β; lanes 2 and 3, two individual recombinant clones.

in an increase of 1340 bp in this fragment and is just detectable in both modified clones (**Fig. 6B**, lanes 2 and 3).

The ε-globin gene is present on a 3751-bp *Eco*RI fragment in pEBAC/148β, whereas the EGFP-Neo/Kan cassette does not have any *Eco*RI sites. The replacement of the ε-globin gene by the EGFP-Neo/Kan cassette should therefore lead to the loss of a 3751-bp band and the appearance of a 5091-bp band after *Eco*RI digestion. High-resolution mapping after *Eco*RI digestion clearly demonstrates the loss of a 3751-bp band with the concomitant appearance of a 5091-bp band in both recombinant clones (**Fig. 7**). At the same time, this gel demonstrates the absence of any minor deletions or rearrangements in the recombinant clones.

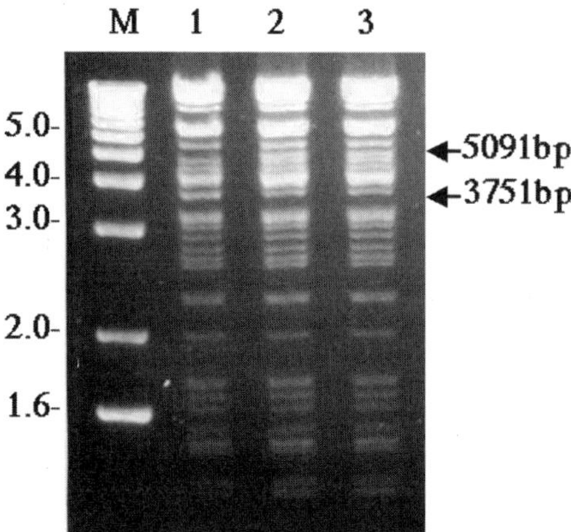

Fig. 7. Restriction enzyme mapping of recombinant clones at high resolution. Two recombinant clones were analyzed at high resolution after digestion with *Eco*RI. A band at 3751 bp disappears in the modified clones and is replaced by a new band at 5091 bp because of the replacement of the ε-globin gene with the EGFP-Neo/Kan cassette. Lane 1, Parent clone, pEBAC/148β; lanes 2 and 3, two individual recombinant clones.

3.4.5. Time Considerations

Fully sequenced PAC or BAC clones can now be identified for most genes of interest from the publicly available genomic libraries. It is advisable before launching into a PAC/BAC modification project to allocate a couple of days for in silico research, to identify all available PAC/BAC clones for the gene of interest and look for unique sequences that may be used as preferred target sites for homologous recombination. If several clones are available, then it may also be better to start with a clone that retains maximum amount of sequence at both the 5' and 3' ends of the genomic locus of interest. It is also necessary at this stage to select restriction nucleases that will allow analysis of the clone before and after modification at medium and high resolution (*Not*I is normally used to release the insert and check for the absence of gross unwanted rearrangements).

The selected genomic clone(s) are streaked out and glycerol stocks are established. The electroporation of the pGETrec plasmid into the selected clone and the preparation of electrocompetent DH10B (BAC, pGETrec) cells retaining both plasmids at high efficiency will require 3–4 d.

The targeting cassette may be prepared and purified in parallel with the cells, so that it can be electroporated into freshly prepared cells.

After electroporation of the targeting cassette into DH10B (BAC, pGETrec) cells, identification of positive recombinants, initially by colony PCR, removal of the pGETrec plasmid, and analysis of miniprep DNA preparations at various levels of resolution and by sequencing will require another 1–2 wk.

Thus, by using the GET Recombination system, it is possible for experienced researchers to complete targeted modifications on BAC clones within 2–3 wk.

4. Concluding Remarks

The GET Recombination system is a powerful tool that can be used to facilitate the introduction of a variety of modifications into genomic fragments in BAC/PAC clones.

In the simplest type of modification, a reporter/antibiotic cassette may be inserted in frame with the gene of interest. The use of the kanamycin/neomycin gene in the cassette facilitates not only the identification of recombinant colonies in bacteria but also the establishment of stably transfected eukaryotic cell lines, thus providing sensitive assays for gene expression under physiologically relevant conditions. Such assays may then be used for the development of high throughput screening assays to facilitate the identification of drugs that may modify gene expression. Insertion of the reporter at the start codon of the gene of interest enables an analysis of the effects of the promoter and other distant regulatory elements and their binding factors, without any interference from downstream elements. However, introduction of the reporter just before the stop codon can allow an analysis of the effects of various intragenic regulatory elements, as well as the impact of various upstream mutations on mRNA processing and gene expression.

The inclusion of an antibiotic selectable marker in the targeting cassette prevents the study of the role of any downstream elements on gene expression. This limitation may be overcome by the in-frame insertion of the reporter gene, without the antibiotic marker. This may be achieved through a two-stage GET Recombination procedure *(11)*, involving first the insertion of a counterselection cassette and then its replacement by the in-frame insertion of the coding sequence of the reporter at both its 5' and 3' ends. The tetracycline gene can be used for both selection and counterselection *(11)*, but its overall efficiency in counterselection is low. However, novel counterselection approaches *(21,22)* may allow such fine modifications to be achieved with much greater efficiency.

Similar approaches may be used to introduce a variety of disease causing mutations or polymorphisms in cloned genomic loci without leaving behind operational sequences *(11,21,22)*. The impact of such modifications may then be studied directly in cell lines and in transgenic animal models. It is anticipated that this approach will be particularly useful in understanding the impact of various polymorphisms on gene expression. Because tens, if not hundreds, of polymorphic differences may exist in any genomic locus from the published sequences, proposals for a direct role of any polymorphism on gene expression should be verified by the insertion of the relevant polymorphism on a fully sequenced clone.

Although there are great hopes for the therapy of a variety of diseases by direct gene delivery, the development of GET Recombination and other similar techniques for the modification and functional analysis of PAC/BAC clones should facilitate the process of drug discovery for the targeted modification of gene expression, so as to overcome or complement the effects of various disease causing mutations.

5. Notes

1. Plasmid pGETrec is available for distribution from Dr P. Ioannou, Cell & Gene Therapy Research Group, Murdoch Childrens Research Institute, Melbourne, Australia (e-mail: ioannoup@cryptic.rch.

unimelb.edu.au) after completion of a material transfer agreement (available for download from the web site: http://murdoch.rch. unimelb.edu.au/pages/lab/cell_gene_therapy/overview.html.)

2. Common antibiotics are used at the final concentrations in media and on plates as follows: ampicillin (Amp), 100 µg/mL; chloramphenicol (Cm), 12.5 µg/mL; or kanamycin (Kan), 25 µg/mL.

3. The use of freshly prepared cells is highly recommended because they exhibit higher transformation efficiencies. The protocol described for the preparation of electrocompetent cells should be performed rapidly whilst taking care to maintain the temperature of the cells and solutions as cold as possible during the procedure. At least four washes with ice cold 10% glycerol should be conducted to ensure that all of the salts present in the culture media have been washed from the cells to avoid arcing of the sample during electroporation.

4. Maximum electroporation efficiency is attained by electroporating 2 µL DNA in 0.5X TE buffer per 40 µL of cells. Using a larger volume of DNA or higher concentration buffer can seriously impact on the electroporation efficiency.

5. It is good to pick a number of independent clones with both plasmids and make miniprep DNA preparations for analysis. Although the pGETrec plasmid is high copy, many of the PAC or BAC bands can still be seen on a gel after digestion with *Eco*RI. This type of analysis may also be used to check the absence of any rearrangements in the PAC/BAC clone prior to GET Recombination.

6. PCR cycling conditions must be determined empirically for each targeting primer pair. In practice, higher yields of good quality PCR product can be obtained using the Expand High Fidelity PCR System (Boehringer Mannheim) and cycling for 10 cycles at the optimum annealing temperature, followed by a further 20 cycles at an annealing temperature of 70°C, which uses the full melting temperature of the long oligos after the initial few cycles have synthesized some full-length template.

7. Often the size of the PCR targeting cassette can comigrate on agarose gels with supercoiled template plasmid DNA. Treatment of the PCR product with *Dpn*I, which selectively digests methylated DNA, destroys template plasmid without affecting the targeting cassette, thus reducing the number of false-positive clones appearing on the plates following GET Recombination.

Acknowledgments

The authors would like to thank Dr Kumaran Narayanan, Dr Mikhail Nefedov, Dr Joe Sarsero, and Lingli Li for their experiences in optimizing the pGETrec system. The work has been supported by a Center Grant from the Thalassaemia International Federation, and the Cyprus Thalassaemia Association, as well as by grants from the NHMRC, the Brockhoff Foundation, the Ronald Geoffrey Arnott Foundation, and the Thalassaemia Societies of New South Wales and Victoria. DJ was partially supported by a Royal Golden Jubilee Scholarship from the Thailand Research Fund.

References

1. Ioannou, P. A., Amemiya, C. T., Garnes, J., Kroisel, P. M., Shizuya, H., Chen, C., et al. (1994) A new bacteriophage P1-derived vector for the propagation of large human DNA fragments. *Nat. Genet.* **6,** 84–89.
2. Shizuya, H., Birren, B., Kim, U. J., Mancino, V., Slepak, T., Tachiiri, Y., et al. (1992) Cloning and stable maintenance of 300-kilobase-pair fragments of human DNA in *Escherichia coli* using an F-factor-based vector. *Proc. Natl. Acad. Sci. USA* **89,** 8794–8797.
3. Yang, X. W., Model, P., and Heintz, N. (1997) Homologous recombination based modification in Escherichia coli and germline transmission in transgenic mice of a bacterial artificial chromosome. *Nat. Biotech.* **15,** 859–865.
4. Murphy, K. C. (1998) Use of bacteriophage lambda recombination functions to promote gene replacement in *Escherichia coli. J. Bacteriol.* **180,** 2063–2071.
5. Zhang, Y., Buchholz, F., Muyrers, J. P., and Stewart, A. F. (1998) A new logic for DNA engineering using recombination in *Escherichia coli. Nat. Genet.* **20,** 123–128.
6. Narayanan, K., Williamson, R., Zhang, Y., Stewart, A. F., and Ioannou, P. A. (1999) Efficient and precise engineering of a 200kb beta-globin human/bacterial artificial chromosome in *E. coli* DH10B using an inducible homologous recombination system. *Gene Ther.* **6,** 442–447.
7. Muyrers, J. P. P., Zhang, Y., Testa, G., and Stewart, A. F. (1999) Rapid modification of bacterial artificial chromosomes by ET-recombination. *Nucleic Acids Res.* **27,** 1555–1557.

8. Yu, D., Ellis, H. M., Lee, E. C., Jenkins, N. A., Copeland, N., and Court, D. L. (2000) An efficient recombination system for chromosome engineering in *Escherichia coli. Proc. Natl. Acad. Sci. USA* **97,** 5978–5983.

9. Ali Imam, A. M., Patrinos, G., De Krom, M., Bottardi, S., Janssens, R. J., Katsantoni, E., et al. (2000) Modification of human β-globin locus PAC clones by homologous recombination in *Escherichia coli. Nucleic Acids Res.* **28,** e65.

10. Orford, M., Nefedov, M., Vadolas, J., Zaibak, F., Williamson, R., and Ioannou P. A. (2000) Engineering EGFP reporter constructs into a 200kb human β-globin BAC clone using GET Recombination. *Nucleic Acids Res.* **28,** e84.

11. Nefedov, M., Williamson, R., and Ioannou P. A. (2000) Insertion of disease-causing mutations in BACs by homologous recombination in Escherichia coli. *Nucleic Acids Res.* **28,** e79.

12. El Karoui, M., Amundsen, S. K., Dabert, P., and Gruss, A. (1999) Gene replacement with linear DNA in electroporated wild-type Escherichia coli. *Nucleic Acids Res.* **27,** 1296–1299.

13. Lee, E. C., Yu, D., De Velasco, J. M., Tessarollo, L., Swing, D. A., Court, D. L., et al. (2001) A highly efficient Escherichia coli-based chromosome engineering system adapted for recombinogenic targeting and subcloning of BAC DNA. *Genomics* **78,** 56–65.

14. Misulovin, Z., Yang, X. W., Yu, W., Heintz, N., and Meffre, E. (2001) A rapid method for targeted modification and screening of recombinant bacterial artificial chromosomes. *J. Immunol. Methods* **257,** 99–105.

15. Zhang, Y., Muyrers, J. P. P., Testa., G., and Stewart, A. F. (2000) DNA cloning by homologous recombination in Escherichia coli. *Nat. Biotech.* **18,** 1314–1317.

16. Swaminathan, S., Ellis, H. M., Waters, L. S., Yu, D., Lee E. C., Court, D. L., et al. (2001) Rapid engineering of bacterial artificial chromosomes using oligonucleotides. *Genesis* **29,** 14–21.

17. Copeland, N., Jenkins, N. A., and Court, D. L. (2001) Recombineering: A powerful new tool for mouse functional genomics. *Nat. Rev. Genet.* **2,** 769–779.

18. Johnson, C. M. and Schleif, R. F. (1995) In vivo induction kinetics of the arabinose promoters in Escherichia coli. *J. Bacteriol.* **177,** 3438–3442.

19. Siegele, D. A. and Hu, J. C. (1997) Gene expression from plasmids containing the araBAD promoter at subsaturating inducer concentra-

tions represents mixed populations. *Proc. Natl. Acad. Sci. USA* **94,** 8168–8172.

20. Laemmli, U. K. (1970) Cleavage of structural proteins during the assembly of the head of bacteriophage T4. *Nature* **227,** 680–685.

21. Jamsai, D., Nefedov, M., Narayanan, K., et al. (2003) Insertion of common mutations into the human beta-globin locus using GET Recombination and an EcoRI endonuclease counterselection cassette. *J. Biotechnol.* **101,** 1–9.

22. Jamsai, D., Orford, M., Nefedov, M., et al. (2003) Targeted modification of a human beta-globin locus BAC clone using GET Recombination and an I-*Sce*I counterselection cassette. *Genomics* **82,** 68–77.

Index